AF761602

mathbasics
(1, 2, 3 ... to Pre-Algebra)

by

Atul Goel

Math Basics (1, 2, 3 ... to Pre-Algebra)

Published by Altus Publishing
www.altuspublishing.com

ISBN: 978-1-61380-061-4
Library of Congress Control Number: 2011902458

Table of Contents

Table of Contents (continued)

Preface

Why Use Math Basics (1, 2, 3 ... to Pre-Algebra)?

Math Basics (1, 2, 3 ... to Pre-Algebra) is designed to help students that are doing poorly in high math classes in college or high school. As students go through elementary and middle school, they learn basic mathematical concepts. However, many times students will have passed their math classes, but not have totally mastered one or more of their basic math skills along the way.

When these students begin higher math (e.g., Algebra) in college or high school, all of a sudden their weaknesses become a stumbling block because higher math requires them to integrate and apply all of the skills from earlier math classes.

A student will likely struggle in higher math unless their specific weaknesses in basic math are identified and fixed. No amount of tutoring in Algebra alone will help the student unless the specific weaknesses in basic math from earlier levels are identified and fixed! For example, a student having trouble solving simultaneous equations in Algebra will not succeed unless you first identify that he really didn't understand signed-number arithmetic from basic math concepts. Or the high school student having difficulty factoring a polynomial may be because he really didn't master exponent multiplication in previous years. Unless you first identify those pre-requisite math skills from earlier years, it would be very difficult helping him improve in higher math classes in college or high school.

So, how do you identify the specific basic math skills from earlier levels that the student is weak in?

This is where our program, Math Basics (1, 2, 3 ... to Pre-Algebra), comes in. We have broken the fundamental math concepts from elementary school through Pre-Algebra into 56 basic units that build upon each other. Using our system, you can specifically identify which basic math skills the student is weak in and give focused help to improve the student's skills. After completing this system, the student should be ready to take on the higher math concepts of college or high school.

How to Use Math Basics (1, 2, 3 ... to Pre-Algebra)

Step 1: Assess the Student's Weaknesses

First you need to determine the specific areas of weakness for the student. To do this, ask the student to take our Assessment Test. Once the student has completed the test, write down which questions the student missed. Each question on the Assessment Test corresponds to a specific Unit number in our series. For example, if the student missed question 14, then he needs to learn the skill associated with Unit 14 of Math Basics (1, 2, 3 ... to Pre-Algebra). By seeing which questions the student misses on our Assessment Test, you can determine precisely which Units in our program he needs to master to get ready for higher math (e.g., Algebra).

Step 2: Start Using the Worksheets

Start with the first question the student missed. Go to Worksheet number for that corresponding Unit and show the student how to do a few of the problems (for example, if the student missed problem #7 on the Assessment Test, go to Unit 7, Worksheet #1). The student can also take that worksheet to his teacher or tutor for guidance. For further practice, keep going through the succeeding worksheets in that Unit until the student can complete all the problems on one worksheet in the time allotted at the top of the worksheet. Again, help the student if he is unable to answer all the problems correctly.

Once the student can complete all the problems on one worksheet correctly within the Target Time specified at the top of each worksheet, he can be considered to have mastered that Unit. Any future updates to our worksheets (if any) will be provided on our website, www.altuspublishing.com.

Once the student has mastered the worksheets from one Unit, move on to the next Unit corresponding to the question the student missed on the Assessment Test. For example, if the student missed questions 7, 24, and 39 on the Assessment Test; after mastering Unit 7, move on to Unit 24.

Work through the system until the student has mastered all the units corresponding to each question missed on the Assessment Test.

Assessment Test

1. How many triangles are there?

▲▲▲▲▲▲▲▲

2. In **2,894,706**, what digit is in the "thousands" place?

3. Round **2,582,569** to the "millions" place.

4. Solve:

$$\begin{array}{r} \mathbf{2} \\ \mathbf{4} \\ + \ \mathbf{3} \\ \hline \end{array}$$

Assessment Test

5. Solve

$$\begin{array}{r} \mathbf{21} \\ \mathbf{13} \\ +\ \mathbf{52} \\ \hline \end{array}$$

6. Solve:

$$\begin{array}{r} \mathbf{56} \\ \mathbf{87} \\ +\ \mathbf{94} \\ \hline \end{array}$$

7. Solve:

$$\begin{array}{r} \mathbf{2972} \\ \mathbf{4799} \\ +\ \mathbf{5064} \\ \hline \end{array}$$

Assessment Test

8. Solve:

$$\begin{array}{r} \mathbf{8} \\ \underline{-\ \mathbf{7}} \end{array}$$

9. Solve:

$$\begin{array}{r} \mathbf{2726} \\ \underline{-\ \mathbf{711}} \end{array}$$

10. Solve:

$$\begin{array}{r} \mathbf{4670} \\ \underline{-\ \mathbf{1352}} \end{array}$$

Assessment Test

11. Solve:

$$\begin{array}{r} \mathbf{8716} \\ -\ \mathbf{4287} \\ \hline \end{array}$$

12. What is: $\mathbf{7 \times 12}$

13. Solve:

$$\begin{array}{r} \mathbf{8} \\ \times\ \mathbf{7} \\ \hline \end{array}$$

14. Solve:

$$\begin{array}{r} \mathbf{79} \\ \times\ \mathbf{4} \\ \hline \end{array}$$

Assessment Test

15. Solve:

$$\begin{array}{r} \mathbf{94} \\ \underline{\times \mathbf{66}} \end{array}$$

16. Solve:

$$\begin{array}{r} \mathbf{622} \\ \underline{\times \mathbf{63}} \end{array}$$

Assessment Test

17. Solve:

$$\begin{array}{r} \mathbf{3797} \\ \times\ \ \mathbf{719} \\ \hline \end{array}$$

18. Solve:

$$3\overline{)8}$$

Assessment Test

20. Solve: $5\overline{)503}$

20. Solve: $36\overline{)8042}$

Assessment Test

21. What is the median of the following numbers:

49, 47, 42, 94

22. Calculate: 5^4

Assessment Test

23. Convert this fraction to a mixed number:

$$\frac{594}{67}$$

24. Convert the mixed number into a simple fraction:

$$16\,\frac{36}{366}$$

Assessment Test

25. Find the prime factors, arrange in order, and express as exponents:

17640

26. Find the GCD of the following numbers:

1715, 48020

Assessment Test

27. Reduce the fraction to lowest terms:

$$\frac{210}{350}$$

28. Solve the problem.
Leave the answer in reduced, simple fraction form:

$$2\frac{39}{11}+8\frac{17}{3}$$

Assessment Test

29. Solve the problem.
Leave the answer in reduced, simple fraction from:

$$6\frac{5}{8} - 2\frac{6}{2}$$

30. Solve the problem.
Leave the answer in reduced, simple fraction form:

$$2\frac{9}{30} + 5\frac{1}{343}$$

Assessment Test

31. Multiply the fractions and reduce to simplest fraction:

$$\frac{10}{3} \times \frac{8}{5}$$

32. Solve the problem and reduce to simplest fraction:

$$\frac{4}{9} + \frac{3}{7}$$

33. Solve the problem and reduce to simplest fraction:

$$\frac{3}{30} \div \frac{375}{18} \div \frac{4}{2}$$

34. Simplify the complex fraction:

$$\frac{2\frac{1}{4} + 1\frac{3}{4} - \frac{3}{4}}{6\frac{1}{2} \div \frac{8}{5} \times 2\frac{1}{6}}$$

Assessment Test

35. Rewrite the expression using the associative law of addition:

    $(56 + 82) + 42$

36. Use the distributive law to rewrite the expression:

    $73 \times (24 + 52)$

Assessment Test

37. Solve: $(2+5)\times 6^2 + 9 \div 3 \times 0$

38. Write the following as a decimal number:

Eighteen hundred-thousandths

Assessment Test

39. Solve the following problem:

4.632 + 56.834918 + 692.083 + 48.638

40. Solve the problem:

7723.77 – 397.68916

Assessment Test

41. Solve:

$$\mathbf{51.01 \times 79.67}$$

42. Solve:

$$\mathbf{2.1 \div 2.4}$$

Assessment Test

43. Convert the following decimal into a fraction:

160.4

44. Convert the following fraction to a decimal:

$$\frac{225}{360}$$

Assessment Test

45. Convert the following percent to a fraction and reduce to lowest terms:

    162.5%

46. Convert the decimal to a percent:

    561.78441

Assessment Test

47. Solve the following problem:

2.8% of 1.9

48. What percent of 180 is 252?

Assessment Test

49. 60 is 75% of what number?

50. For every 5 computers sold, 3 people buy speakers. If 27 people buy speakers, how many computers were sold?

Assessment Test

51. If an 8 gallon fish tank holds 64 small fish, then how many gallon tank is needed to hold 48 small fish?

52. Solve the following problem:

    -52 + 48

Assessment Test

53. Solve the following problem:

-58 – (-70)

54. Solve the following problem:

-61 × -51

Assessment Test

55. Solve the following problem:

$$124 \div -62$$

56. Solve the following problem:

$$|328 \div -82|$$

Counting Numbers

mathbasics (1,2,3 ... to Pre-Algebra)

Unit: 01

Worksheet: I

Target time: 1 min Student's time: min

Count the items:

______

______

______

______

______

______

______

______

______

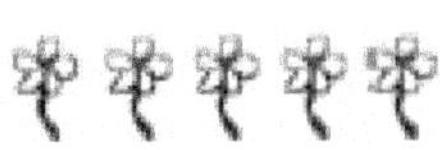

Draw the number of items indicated:

1

3

5

2

6

4

8

9

7

Counting Numbers

Unit: 01
Worksheet: 2

mathbasics (1,2,3 ... to Pre-Algebra)   Target time: **1** min   Student's time: ____ min

Count the items:

____

____

____

____

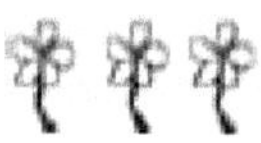

____

____

____

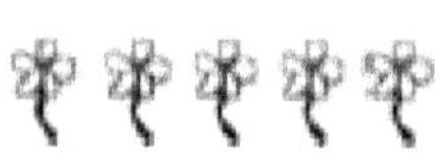

____

____

Draw the number of items indicated:

8

6

2

5

9

1

3

4

7

mathbasics (1,2,3 ... to Pre-Algebra) Target time: 1 min Student's time: min

Count the items:

_____

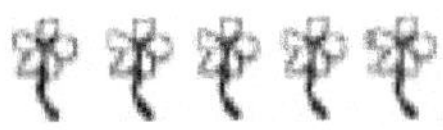

_____

_____

_____

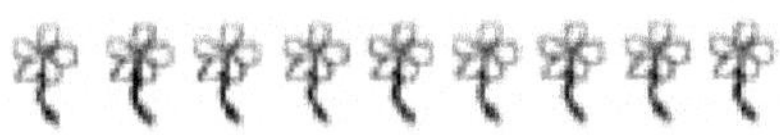

_____

_____

_____

_____

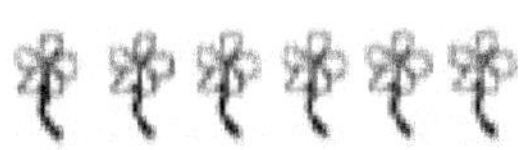

_____

Draw the number of items indicated:

3

1

6

8

4

7

5

9

2

mathbasics (1,2,3 ... to Pre-Algebra) Target time: 5 min Student's time: ____ min

1. In 8,579,643, what is the place value of:
   - 7 -
   - 4 -
   - 8 -
   - 3 -
   - 9 -
   - 5 -
   - 6 -

2. In 2,810,596, what is the place value of:
   - 8 -
   - 1 -
   - 9 -
   - 5 -
   - 6 -
   - 2 -
   - 0 -

3. In 7,125,389, what is the place value of:
   - 1 -
   - 8 -
   - 3 -
   - 7 -
   - 2 -
   - 5 -
   - 9 -

4. In 6,349,582, what digit is in the:
   - ones place -
   - tens place -
   - millions place -
   - hundreds place -
   - thousands place -
   - ten-thousands place -
   - hundred-thousands place -

5. In 9,265,341, what digit is in the:
   - millions place -
   - tens place -
   - hundred-thousands place -
   - hundreds place -
   - ones place -
   - ten-thousands place -
   - thousands place -

6. In 8,192,057, what digit is in the:
   - tens place -
   - thousands place -
   - millions place -
   - ones place -
   - hundreds place -
   - hundred-thousands place -
   - ten-thousands place -

mathbasics (1,2,3 ... to Pre-Algebra) **Target time: 5 min** **Student's time: ____ min**

1. In 3,865,290, what is the place value of:
   - 5 -
   - 3 -
   - 6 -
   - 0 -
   - 8 -
   - 9 -
   - 2 -

2. In 7,162,489, what is the place value of:
   - 2 -
   - 8 -
   - 1 -
   - 6 -
   - 7 -
   - 9 -
   - 4 -

3. In 9,702,483, what is the place value of:
   - 0 -
   - 7 -
   - 4 -
   - 9 -
   - 2 -
   - 8 -
   - 3 -

4. In 5,726,983, what digit is in the:
   - tens place -
   - ten-thousands place -
   - ones place -
   - millions place -
   - hundreds place -
   - hundred-thousands place -
   - thousands place -

5. In 2,846,971, what digit is in the:
   - ten-thousands place -
   - ones place -
   - millions place -
   - thousands place -
   - hundreds place -
   - hundred-thousands place -
   - tens place -

6. In 6,382,147, what digit is in the:
   - thousands place -
   - hundred-thousands place -
   - tens place -
   - millions place -
   - hundreds place -
   - ones place -
   - ten-thousands place -

mathbasics (1,2,3 ... to Pre-Algebra) Target time: 5 min Student's time: ____ min

1. In 8,214,063, what is the place value of:
   2 -
   1 -
   8 -
   4 -
   3 -
   0 -
   6 -

2. In 8,309,421, what is the place value of:
   2 -
   3 -
   4 -
   9 -
   0 -
   8 -
   1 -

3. In 1,832,904, what is the place value of:
   4 -
   9 -
   0 -
   1 -
   3 -
   8 -
   2 -

4. In 9,645,378, what digit is in the:
   ten-thousands place -
   hundred-thousands place -
   tens place -
   thousands place -
   hundreds place -
   ones place -
   millions place -

5. In 5,023,748, what digit is in the:
   hundreds place -
   ten-thousands place -
   tens place -
   ones place -
   hundred-thousands place -
   thousands place -
   millions place -

6. In 4,581,739, what digit is in the:
   millions place -
   ones place -
   tens place -
   thousands place -
   ten-thousands place -
   hundreds place -
   hundred-thousands place -

Numbers 1-100 and Rounding

Unit: 03
Worksheet: 1

mathbasics (1,2,3 ... to Pre-Algebra) **Target time: 5 min** **Student's time: ____ min**

Write each number in word form:

28 -	40 -	14 -
15 -	2 -	81 -
20 -	50 -	22 -
71 -	25 -	39 -
9 -	35 -	75 -
48 -	45 -	16 -
37 -	58 -	7 -
5 -	6 -	42 -
60 -	84 -	30 -
93 -	67 -	31 -

Round each number to the designated place value:

9,852,702 (to ten-thousands) -	3,773,527 (to hundred-thousands) -
1,787,191 (to hundreds) -	7,706,633 (to thousands) -
8,782,124 (to millions) -	5,225,858 (to thousands) -
7,954,439 (to millions) -	7,747,684 (to tens) -

Write each number in word form:

70 -	79 -	38 -
75 -	51 -	59 -
65 -	8 -	78 -
15 -	99 -	9 -
23 -	86 -	67 -
54 -	4 -	2 -
49 -	96 -	61 -
98 -	10 -	88 -
3 -	17 -	47 -
89 -	27 -	55 -

Round each number to the designated place value:

2,722,431 (to hundred-thousands) -	3,565,787 (to millions) -
3,609,912 (to hundreds) -	116,087 (to tens) -
4,245,029 (to ten-thousands) -	3,602,339 (to millions) -
8,382,844 (to thousands) -	3,816,726 (to thousands) -

Numbers 1-100 and Rounding

mathbasics (1,2,3 ... to Pre-Algebra) Target time: 5 min Student's time: ____ min

Write each number in word form:

29 -	96 -	33 -
46 -	34 -	41 -
20 -	91 -	95 -
50 -	77 -	13 -
40 -	22 -	12 -
1 -	48 -	25 -
5 -	18 -	17 -
52 -	24 -	57 -
45 -	37 -	16 -
36 -	53 -	31 -

Round each number to the designated place value:

6,226,440 (to hundred-thousands) -	7,906,701 (to tens) -
746,979 (to tens) -	3,999,500 (to hundreds) -
1,823,539 (to millions) -	7,870,323 (to ten-thousands) -
8,608,462 (to hundred-thousands) -	4,097,709 (to thousands) -

Target time: 1 min **Student's time: ____ min**

Solve the following problems:

$$\begin{array}{r} 4 \\ +2 \\ \hline \end{array} \qquad \begin{array}{r} 4 \\ +4 \\ \hline \end{array} \qquad \begin{array}{r} 4 \\ +5 \\ \hline \end{array}$$

$$\begin{array}{r} 3 \\ +1 \\ \hline \end{array} \qquad \begin{array}{r} 7 \\ +1 \\ \hline \end{array} \qquad \begin{array}{r} 2 \\ +7 \\ \hline \end{array}$$

$$\begin{array}{r} 1 \\ +6 \\ \hline \end{array} \qquad \begin{array}{r} 3 \\ +1 \\ \hline \end{array} \qquad \begin{array}{r} 4 \\ +1 \\ \hline \end{array}$$

$$\begin{array}{r} 6 \\ 1 \\ +2 \\ \hline \end{array} \qquad \begin{array}{r} 6 \\ 2 \\ +1 \\ \hline \end{array} \qquad \begin{array}{r} 1 \\ 7 \\ +1 \\ \hline \end{array}$$

$$\begin{array}{r} 6 \\ 1 \\ +1 \\ \hline \end{array} \qquad \begin{array}{r} 2 \\ 4 \\ +3 \\ \hline \end{array} \qquad \begin{array}{r} 6 \\ 1 \\ +1 \\ \hline \end{array}$$

Addition, Level 1

Unit: 04
Worksheet: 2

mathbasics (1,2,3 ... to Pre-Algebra) **Target time: 1 min** **Student's time: ____ min**

Solve the following problems:

$$\begin{array}{r} 5 \\ +3 \\ \hline \end{array} \qquad \begin{array}{r} 2 \\ +3 \\ \hline \end{array} \qquad \begin{array}{r} 1 \\ +4 \\ \hline \end{array}$$

$$\begin{array}{r} 4 \\ +1 \\ \hline \end{array} \qquad \begin{array}{r} 2 \\ +5 \\ \hline \end{array} \qquad \begin{array}{r} 2 \\ +4 \\ \hline \end{array}$$

$$\begin{array}{r} 7 \\ +1 \\ \hline \end{array} \qquad \begin{array}{r} 1 \\ +3 \\ \hline \end{array} \qquad \begin{array}{r} 7 \\ +2 \\ \hline \end{array}$$

$$\begin{array}{r} 7 \\ 1 \\ +1 \\ \hline \end{array} \qquad \begin{array}{r} 1 \\ 6 \\ +2 \\ \hline \end{array} \qquad \begin{array}{r} 4 \\ 1 \\ +2 \\ \hline \end{array}$$

$$\begin{array}{r} 3 \\ 5 \\ +1 \\ \hline \end{array} \qquad \begin{array}{r} 4 \\ 2 \\ +1 \\ \hline \end{array} \qquad \begin{array}{r} 5 \\ 3 \\ +1 \\ \hline \end{array}$$

Addition, Level 1

Unit: 04
Worksheet: 3

mathbasics (1,2,3 ... to Pre-Algebra) **Target time: 1 min** **Student's time: ____ min**

Solve the following problems:

$$\begin{array}{r} 1 \\ +3 \\ \hline \end{array} \qquad \begin{array}{r} 3 \\ +2 \\ \hline \end{array} \qquad \begin{array}{r} 8 \\ +1 \\ \hline \end{array}$$

$$\begin{array}{r} 7 \\ +2 \\ \hline \end{array} \qquad \begin{array}{r} 5 \\ +4 \\ \hline \end{array} \qquad \begin{array}{r} 8 \\ +1 \\ \hline \end{array}$$

$$\begin{array}{r} 4 \\ +5 \\ \hline \end{array} \qquad \begin{array}{r} 4 \\ +5 \\ \hline \end{array} \qquad \begin{array}{r} 1 \\ +2 \\ \hline \end{array}$$

$$\begin{array}{r} 4 \\ 3 \\ +1 \\ \hline \end{array} \qquad \begin{array}{r} 6 \\ 2 \\ +1 \\ \hline \end{array} \qquad \begin{array}{r} 3 \\ 2 \\ +3 \\ \hline \end{array}$$

$$\begin{array}{r} 4 \\ 1 \\ +3 \\ \hline \end{array} \qquad \begin{array}{r} 5 \\ 1 \\ +3 \\ \hline \end{array} \qquad \begin{array}{r} 6 \\ 1 \\ +2 \\ \hline \end{array}$$

Addition, Level 2

Unit: 05
Worksheet: 1

mathbasics (1,2,3 ... to Pre-Algebra)

Target time: 1 min   Student's time: ____ min

Solve the following problems:

$$\begin{array}{r} 63 \\ +11 \\ \hline \end{array} \qquad \begin{array}{r} 11 \\ +71 \\ \hline \end{array} \qquad \begin{array}{r} 20 \\ +40 \\ \hline \end{array}$$

$$\begin{array}{r} 58 \\ +20 \\ \hline \end{array} \qquad \begin{array}{r} 77 \\ +10 \\ \hline \end{array} \qquad \begin{array}{r} 25 \\ +51 \\ \hline \end{array}$$

$$\begin{array}{r} 17 \\ +51 \\ \hline \end{array} \qquad \begin{array}{r} 77 \\ +12 \\ \hline \end{array} \qquad \begin{array}{r} 11 \\ +12 \\ \hline \end{array}$$

$$\begin{array}{r} 13 \\ 21 \\ +52 \\ \hline \end{array} \qquad \begin{array}{r} 38 \\ 20 \\ +31 \\ \hline \end{array} \qquad \begin{array}{r} 41 \\ 26 \\ +12 \\ \hline \end{array}$$

$$\begin{array}{r} 73 \\ 15 \\ +10 \\ \hline \end{array} \qquad \begin{array}{r} 61 \\ 18 \\ +10 \\ \hline \end{array} \qquad \begin{array}{r} 24 \\ 23 \\ +30 \\ \hline \end{array}$$

Addition, Level 2

Unit: 05
Worksheet: 2

mathbasics (1,2,3 ... to Pre-Algebra) **Target time: 1 min** **Student's time: ____ min**

Solve the following problems:

$\begin{array}{r} 16 \\ +63 \\ \hline \end{array}$	$\begin{array}{r} 77 \\ +11 \\ \hline \end{array}$	$\begin{array}{r} 31 \\ +54 \\ \hline \end{array}$
$\begin{array}{r} 55 \\ +14 \\ \hline \end{array}$	$\begin{array}{r} 72 \\ +15 \\ \hline \end{array}$	$\begin{array}{r} 26 \\ +31 \\ \hline \end{array}$
$\begin{array}{r} 55 \\ +30 \\ \hline \end{array}$	$\begin{array}{r} 44 \\ +13 \\ \hline \end{array}$	$\begin{array}{r} 64 \\ +14 \\ \hline \end{array}$
$\begin{array}{r} 22 \\ 24 \\ +11 \\ \hline \end{array}$	$\begin{array}{r} 15 \\ 63 \\ +20 \\ \hline \end{array}$	$\begin{array}{r} 51 \\ 17 \\ +21 \\ \hline \end{array}$
$\begin{array}{r} 27 \\ 12 \\ +20 \\ \hline \end{array}$	$\begin{array}{r} 67 \\ 21 \\ +10 \\ \hline \end{array}$	$\begin{array}{r} 33 \\ 54 \\ +11 \\ \hline \end{array}$

Addition, Level 2

Unit: 05
Worksheet: 3

mathbasics (1,2,3 ... to Pre-Algebra) **Target time: 1 min** **Student's time: ____ min**

Solve the following problems:

12 +55	21 +67	76 +11
25 +54	42 +42	58 +30
14 +22	78 +10	47 +12
16 70 +10	40 21 +24	17 52 +20
29 40 +10	48 40 +11	74 13 +12

Addition, Level 3

Unit: 06
Worksheet: 1

mathbasics (1,2,3 ... to Pre-Algebra) **Target time: 2 min** **Student's time: ____ min**

Solve the following problems:

39	33	94
+ 68	+ 99	+ 66

58	57	15
+ 67	+ 54	+ 49

61	73	28
+ 29	+ 57	+ 25

22	23	69
18	29	76
+ 85	+ 17	+ 24

56	24	75
87	27	69
+ 94	+ 52	+ 72

Addition, Level 3

Unit: 06
Worksheet: 2

mathbasics (1,2,3 ... to Pre-Algebra) **Target time: 2 min** **Student's time: ____ min**

Solve the following problems:

$$\begin{array}{r} 45 \\ +98 \\ \hline \end{array} \qquad \begin{array}{r} 84 \\ +57 \\ \hline \end{array} \qquad \begin{array}{r} 41 \\ +39 \\ \hline \end{array}$$

$$\begin{array}{r} 84 \\ +57 \\ \hline \end{array} \qquad \begin{array}{r} 28 \\ +34 \\ \hline \end{array} \qquad \begin{array}{r} 66 \\ +24 \\ \hline \end{array}$$

$$\begin{array}{r} 67 \\ +33 \\ \hline \end{array} \qquad \begin{array}{r} 33 \\ +59 \\ \hline \end{array} \qquad \begin{array}{r} 44 \\ +87 \\ \hline \end{array}$$

$$\begin{array}{r} 47 \\ 78 \\ +27 \\ \hline \end{array} \qquad \begin{array}{r} 33 \\ 99 \\ +44 \\ \hline \end{array} \qquad \begin{array}{r} 17 \\ 27 \\ +63 \\ \hline \end{array}$$

$$\begin{array}{r} 76 \\ 16 \\ +28 \\ \hline \end{array} \qquad \begin{array}{r} 96 \\ 45 \\ +49 \\ \hline \end{array} \qquad \begin{array}{r} 67 \\ 75 \\ +58 \\ \hline \end{array}$$

Addition, Level 3

Unit: 06
Worksheet: 3

mathbasics (1,2,3 ... to Pre-Algebra) **Target time: 2 min** **Student's time: ____ min**

Solve the following problems:

$$\begin{array}{r}28\\+85\\\hline\end{array}\qquad\begin{array}{r}68\\+27\\\hline\end{array}\qquad\begin{array}{r}53\\+57\\\hline\end{array}$$

$$\begin{array}{r}93\\+98\\\hline\end{array}\qquad\begin{array}{r}45\\+46\\\hline\end{array}\qquad\begin{array}{r}96\\+\ \ 5\\\hline\end{array}$$

$$\begin{array}{r}91\\+29\\\hline\end{array}\qquad\begin{array}{r}65\\+58\\\hline\end{array}\qquad\begin{array}{r}42\\+38\\\hline\end{array}$$

$$\begin{array}{r}67\\04\\+60\\\hline\end{array}\qquad\begin{array}{r}31\\59\\+14\\\hline\end{array}\qquad\begin{array}{r}17\\59\\+64\\\hline\end{array}$$

$$\begin{array}{r}37\\54\\+23\\\hline\end{array}\qquad\begin{array}{r}82\\78\\+23\\\hline\end{array}\qquad\begin{array}{r}34\\89\\+10\\\hline\end{array}$$

Addition, Level 4

Unit: 07
Worksheet: 1

mathbasics (1,2,3 ... to Pre-Algebra)

Target time: 4 min **Student's time: ____ min**

Solve the following problems:

9499 + 2546	1911 + 9489	7333 + 5609
5448 + 6736	2345 + 9805	7207 + 6985
3168 + 7664	1117 + 8523	9658 + 4583
8995 9886 + 9407	1702 8289 + 8087	7362 9549 + 5315
2972 4799 + 5064	3501 5989 + 2442	9894 1197 + 645

mathbasics (1,2,3 ... to Pre-Algebra) Target time: 4 min Student's time: ____ min

Solve the following problems:

4309 +3726	3378 +2506	9579 +3635
7288 +8334	7321 +1479	8412 +268
7492 +369	6465 +8077	4965 +4486
9049 2101 +5916	3849 2032 +165	2306 7897 +8452
6513 5389 +1275	4155 3297 +5913	8187 6484 +46

Addition, Level 4

Unit: 07
Worksheet: 3

mathbasics (1,2,3 ... to Pre-Algebra) **Target time: 4 min** **Student's time: ____ min**

Solve the following problems:

$\begin{array}{r} 4345 \\ +6105 \\ \hline \end{array}$	$\begin{array}{r} 3546 \\ +8416 \\ \hline \end{array}$	$\begin{array}{r} 4263 \\ +4597 \\ \hline \end{array}$
$\begin{array}{r} 9739 \\ +7852 \\ \hline \end{array}$	$\begin{array}{r} 2978 \\ +1785 \\ \hline \end{array}$	$\begin{array}{r} 3302 \\ +7248 \\ \hline \end{array}$
$\begin{array}{r} 5149 \\ +6396 \\ \hline \end{array}$	$\begin{array}{r} 1258 \\ +4 \\ \hline \end{array}$	$\begin{array}{r} 9711 \\ +4949 \\ \hline \end{array}$
$\begin{array}{r} 3581 \\ 7329 \\ +4711 \\ \hline \end{array}$	$\begin{array}{r} 1162 \\ 4208 \\ +2072 \\ \hline \end{array}$	$\begin{array}{r} 7822 \\ 8798 \\ +2721 \\ \hline \end{array}$
$\begin{array}{r} 2436 \\ 1475 \\ +5439 \\ \hline \end{array}$	$\begin{array}{r} 6348 \\ 4203 \\ +1239 \\ \hline \end{array}$	$\begin{array}{r} 9303 \\ 1237 \\ +7689 \\ \hline \end{array}$

Subtraction, Level 1

Unit: 08
Worksheet: 1

mathbasics (1,2,3 ... to Pre-Algebra) **Target time: 1 min** **Student's time: ____ min**

Solve the following problems:

9 - 5	6 - 2	1 - 1
4 - 3	4 - 4	3 - 2
9 - 7	5 - 5	7 - 3
9 - 5	3 - 1	7 - 6
2 - 1	3 - 3	7 - 4

Subtraction, Level 1

Unit: 08
Worksheet: 2

mathbasics (1,2,3 ... to Pre-Algebra)   **Target time: 1 min**   **Student's time: ____ min**

Solve the following problems:

$\begin{array}{r} 2 \\ -\ 2 \\ \hline \end{array}$	$\begin{array}{r} 9 \\ -\ 8 \\ \hline \end{array}$	$\begin{array}{r} 8 \\ -\ 1 \\ \hline \end{array}$
$\begin{array}{r} 3 \\ -\ 1 \\ \hline \end{array}$	$\begin{array}{r} 6 \\ -\ 5 \\ \hline \end{array}$	$\begin{array}{r} 2 \\ -\ 2 \\ \hline \end{array}$
$\begin{array}{r} 2 \\ -\ 1 \\ \hline \end{array}$	$\begin{array}{r} 6 \\ -\ 4 \\ \hline \end{array}$	$\begin{array}{r} 9 \\ -\ 4 \\ \hline \end{array}$
$\begin{array}{r} 9 \\ -\ 9 \\ \hline \end{array}$	$\begin{array}{r} 8 \\ -\ 2 \\ \hline \end{array}$	$\begin{array}{r} 2 \\ -\ 2 \\ \hline \end{array}$
$\begin{array}{r} 9 \\ -\ 3 \\ \hline \end{array}$	$\begin{array}{r} 8 \\ -\ 2 \\ \hline \end{array}$	$\begin{array}{r} 3 \\ -\ 2 \\ \hline \end{array}$

Subtraction, Level 1

Unit: 08
Worksheet: 3

mathbasics (1,2,3 ... to Pre-Algebra)   **Target time: 1 min   Student's time: ____ min**

Solve the following problems:

$\begin{array}{r} 7 \\ -4 \\ \hline \end{array}$	$\begin{array}{r} 2 \\ -2 \\ \hline \end{array}$	$\begin{array}{r} 3 \\ -1 \\ \hline \end{array}$
$\begin{array}{r} 7 \\ -5 \\ \hline \end{array}$	$\begin{array}{r} 3 \\ -2 \\ \hline \end{array}$	$\begin{array}{r} 3 \\ -2 \\ \hline \end{array}$
$\begin{array}{r} 3 \\ -2 \\ \hline \end{array}$	$\begin{array}{r} 5 \\ -5 \\ \hline \end{array}$	$\begin{array}{r} 1 \\ -1 \\ \hline \end{array}$
$\begin{array}{r} 3 \\ -3 \\ \hline \end{array}$	$\begin{array}{r} 5 \\ -1 \\ \hline \end{array}$	$\begin{array}{r} 2 \\ -2 \\ \hline \end{array}$
$\begin{array}{r} 7 \\ -2 \\ \hline \end{array}$	$\begin{array}{r} 9 \\ -5 \\ \hline \end{array}$	$\begin{array}{r} 4 \\ -1 \\ \hline \end{array}$

Subtraction, Level 2

Unit: 09
Worksheet: 1

mathbasics (1,2,3 ... to Pre-Algebra) **Target time: 2 min** **Student's time: ____ min**

Solve the following problems:

5814 - 903	5427 - 3901	9112 - 3
9365 - 8182	1474 - 49	7545 - 7365
2858 - 1649	8771 - 3191	6352 - 2081
1719 - 913	7646 - 7209	8115 - 6108
7443 - 272	1587 - 656	4679 - 1289

Subtraction, Level 2

Unit: 09
Worksheet: 2

mathbasics (1,2,3 ... to Pre-Algebra) **Target time: 2 min Student's time: ____ min**

Solve the following problems:

6646 - 716	6614 - 901	1743 - 813
4665 - 3493	8175 - 814	756 - 83
4086 - 2068	5085 - 3059	5281 - 5191
5561 - 4721	9161 - 7116	7611 - 6711
1525 - 614	8236 - 5616	8434 - 2074

Subtraction, Level 2

Unit: 09
Worksheet: 3

mathbasics (1,2,3 ... to Pre-Algebra) **Target time: 2 min** **Student's time: ____ min**

Solve the following problems:

9927 - 847	2825 - 925	9177 - 4404
7697 - 6678	9977 - 2258	7647 - 6481
7816 - 7571	4451 - 1361	3502 - 1902
1128 - 728	769 - 483	5144 - 5008
6546 - 1228	9345 - 9092	6392 - 3029

Subtraction, Level 3

Unit: 10
Worksheet: 1

mathbasics (1,2,3 ... to Pre-Algebra)

Target time: 3 min **Student's time: ____ min**

Solve the following problems:

4906 - 2751	7076 - 2211	4704 - 4521
5703 - 1173	2065 - 1501	1990 - 1706
7709 - 2674	2036 - 611	2101 - 1031
1072 - 441	3105 - 3013	660 - 403
8050 - 8041	7038 - 4831	4670 - 1352

Subtraction, Level 3

Unit: 10
Worksheet: 2

mathbasics (1,2,3 ... to Pre-Algebra) **Target time: 3 min** **Student's time: ____ min**

Solve the following problems:

3570 - 1149	4850 - 1406	2940 - 13
770 - 714	7502 - 3492	5840 - 404
3207 - 3162	8905 - 3463	7040 - 4015
2130 - 2105	9076 - 3401	2054 - 1921
4280 - 4165	3013 - 2501	6072 - 4821

Subtraction, Level 3

Unit: 10
Worksheet: 3

mathbasics (1,2,3 ... to Pre-Algebra) **Target time: 3 min** **Student's time: ____ min**

Solve the following problems:

5103 - 11	1820 - 314	8097 - 6511
7240 - 7035	1790 - 1025	6260 - 2239
2022 - 201	8906 - 6345	4880 - 228
7403 - 3031	8708 - 4184	9302 - 3232
4049 - 1711	6520 - 4105	3807 - 3036

Subtraction, Level 4

Unit: 11
Worksheet: 1

mathbasics (1,2,3 ... to Pre-Algebra) **Target time: 4 min** **Student's time: ____ min**

Solve the following problems:

9051 - 8671	4419 - 2596	3502 - 2891
4587 - 2796	3960 - 397	9642 - 7566
6426 - 3975	2178 - 1598	3848 - 2977
955 - 578	4404 - 2551	8716 - 4287
8156 - 1068	8671 - 7891	8531 - 2861

Subtraction, Level 4

Unit: 11
Worksheet: 2

mathbasics (1,2,3 ... to Pre-Algebra) **Target time: 4 min** **Student's time: ____ min**

Solve the following problems:

5551 - 3971	7320 - 2235	4251 - 1881
4321 - 961	2585 - 1397	3716 - 2826
5813 - 921	1337 - 444	8932 - 2848
5584 - 3396	3404 - 912	5564 - 3672
1755 - 276	3533 - 1296	3860 - 491

Subtraction, Level 4

Unit: 11
Worksheet: 3

mathbasics (1,2,3 ... to Pre-Algebra) **Target time: 4 min** **Student's time: ____ min**

Solve the following problems:

6850 - 2693	9174 - 6692	7544 - 2963
452 - 185	3165 - 1076	3755 - 1993
5707 - 2957	9488 - 1099	9565 - 3695
2512 - 356	3705 - 3049	1170 - 84
6404 - 2721	6843 - 5375	1626 - 773

Multiplication, Level 1

Unit: 12
Worksheet: 1

mathbasics (1,2,3 ... to Pre-Algebra) **Target time: 1 min** **Student's time: ____ min**

Solve the following problems:

$5 \times 7 =$	$10 \times 9 =$	$8 \times 3 =$
$8 \times 8 =$	$2 \times 8 =$	$7 \times 1 =$
$8 \times 12 =$	$7 \times 1 =$	$7 \times 12 =$
$8 \times 6 =$	$10 \times 4 =$	$3 \times 9 =$
$2 \times 2 =$	$5 \times 6 =$	$4 \times 9 =$
$6 \times 9 =$	$12 \times 10 =$	$9 \times 2 =$
$7 \times 10 =$	$8 \times 12 =$	$10 \times 11 =$
$4 \times 3 =$	$6 \times 10 =$	$9 \times 8 =$
$12 \times 12 =$	$9 \times 3 =$	$5 \times 2 =$
$3 \times 4 =$	$7 \times 3 =$	$4 \times 5 =$
$10 \times 7 =$	$12 \times 2 =$	$3 \times 5 =$
$12 \times 6 =$	$12 \times 2 =$	$8 \times 10 =$

Multiplication, Level 1

Unit: 12
Worksheet: 2

mathbasics (1,2,3 ... to Pre-Algebra) **Target time: 1 min** **Student's time: ____ min**

Solve the following problems:

$4 \times 9 =$	$7 \times 3 =$	$3 \times 5 =$
$12 \times 1 =$	$9 \times 3 =$	$8 \times 11 =$
$5 \times 11 =$	$9 \times 8 =$	$1 \times 4 =$
$7 \times 7 =$	$9 \times 11 =$	$9 \times 4 =$
$7 \times 12 =$	$4 \times 10 =$	$10 \times 7 =$
$9 \times 7 =$	$2 \times 12 =$	$12 \times 2 =$
$8 \times 11 =$	$7 \times 1 =$	$3 \times 5 =$
$11 \times 7 =$	$9 \times 10 =$	$11 \times 3 =$
$1 \times 1 =$	$9 \times 1 =$	$9 \times 9 =$
$10 \times 9 =$	$5 \times 6 =$	$6 \times 1 =$
$12 \times 5 =$	$1 \times 10 =$	$8 \times 4 =$
$2 \times 10 =$	$10 \times 9 =$	$6 \times 1 =$

Multiplication, Level 1

Unit: 12
Worksheet: 3

mathbasics (1,2,3 ... to Pre-Algebra) **Target time: 1 min** **Student's time: ____ min**

Solve the following problems:

$11 \times 4 =$	$4 \times 3 =$	$5 \times 3 =$
$11 \times 8 =$	$12 \times 5 =$	$7 \times 12 =$
$11 \times 7 =$	$5 \times 12 =$	$1 \times 6 =$
$1 \times 4 =$	$4 \times 9 =$	$9 \times 2 =$
$4 \times 7 =$	$10 \times 6 =$	$6 \times 2 =$
$12 \times 7 =$	$3 \times 9 =$	$6 \times 11 =$
$2 \times 5 =$	$11 \times 4 =$	$2 \times 2 =$
$8 \times 4 =$	$1 \times 8 =$	$8 \times 6 =$
$1 \times 11 =$	$8 \times 8 =$	$9 \times 12 =$
$12 \times 1 =$	$5 \times 7 =$	$6 \times 12 =$
$8 \times 5 =$	$2 \times 3 =$	$6 \times 5 =$
$12 \times 8 =$	$12 \times 7 =$	$12 \times 10 =$

Multiplication, Level 2

Unit: 13
Worksheet: 1

mathbasics (1,2,3 ... to Pre-Algebra) **Target time: 1 min** **Student's time: ____ min**

Solve the following problems:

5×1	6×9	4×6
5×5	8×3	7×9
8×7	9×5	2×8
4×5	5×5	4×5
5×3	2×2	7×2

Multiplication, Level 2

Unit: 13
Worksheet: 2

mathbasics (1,2,3 ... to Pre-Algebra) **Target time: 1 min** **Student's time: ____ min**

Solve the following problems:

7×2	2×3	2×6
9×7	2×7	9×8
7×8	6×6	3×7
5×6	3×1	2×6
5×6	9×7	7×6

mathbasics (1,2,3 ... to Pre-Algebra) **Target time: 1 min Student's time: ____ min**

Solve the following problems:

$\begin{array}{r} 8 \\ \times 4 \\ \hline \end{array}$ $\begin{array}{r} 8 \\ \times 9 \\ \hline \end{array}$ $\begin{array}{r} 7 \\ \times 1 \\ \hline \end{array}$

$\begin{array}{r} 6 \\ \times 6 \\ \hline \end{array}$ $\begin{array}{r} 7 \\ \times 8 \\ \hline \end{array}$ $\begin{array}{r} 4 \\ \times 7 \\ \hline \end{array}$

$\begin{array}{r} 6 \\ \times 1 \\ \hline \end{array}$ $\begin{array}{r} 5 \\ \times 3 \\ \hline \end{array}$ $\begin{array}{r} 7 \\ \times 8 \\ \hline \end{array}$

$\begin{array}{r} 9 \\ \times 2 \\ \hline \end{array}$ $\begin{array}{r} 5 \\ \times 3 \\ \hline \end{array}$ $\begin{array}{r} 2 \\ \times 6 \\ \hline \end{array}$

$\begin{array}{r} 8 \\ \times 7 \\ \hline \end{array}$ $\begin{array}{r} 3 \\ \times 8 \\ \hline \end{array}$ $\begin{array}{r} 4 \\ \times 1 \\ \hline \end{array}$

Solve the following problems:

$\begin{array}{r} 67 \\ \times 3 \\ \hline \end{array}$	$\begin{array}{r} 59 \\ \times 5 \\ \hline \end{array}$	$\begin{array}{r} 81 \\ \times 3 \\ \hline \end{array}$
$\begin{array}{r} 83 \\ \times 1 \\ \hline \end{array}$	$\begin{array}{r} 18 \\ \times 5 \\ \hline \end{array}$	$\begin{array}{r} 82 \\ \times 9 \\ \hline \end{array}$
$\begin{array}{r} 15 \\ \times 7 \\ \hline \end{array}$	$\begin{array}{r} 47 \\ \times 3 \\ \hline \end{array}$	$\begin{array}{r} 39 \\ \times 5 \\ \hline \end{array}$
$\begin{array}{r} 45 \\ \times 7 \\ \hline \end{array}$	$\begin{array}{r} 85 \\ \times 5 \\ \hline \end{array}$	$\begin{array}{r} 37 \\ \times 2 \\ \hline \end{array}$
$\begin{array}{r} 79 \\ \times 4 \\ \hline \end{array}$	$\begin{array}{r} 55 \\ \times 5 \\ \hline \end{array}$	$\begin{array}{r} 74 \\ \times 1 \\ \hline \end{array}$

Multiplication, Level 3

Unit: 14
Worksheet: 2

mathbasics (1,2,3 ... to Pre-Algebra) Target time: 2 min Student's time: ____ min

Solve the following problems:

$$\begin{array}{r} 13 \\ \times 4 \\ \hline \end{array} \qquad \begin{array}{r} 37 \\ \times 6 \\ \hline \end{array} \qquad \begin{array}{r} 91 \\ \times 1 \\ \hline \end{array}$$

$$\begin{array}{r} 91 \\ \times 7 \\ \hline \end{array} \qquad \begin{array}{r} 22 \\ \times 9 \\ \hline \end{array} \qquad \begin{array}{r} 32 \\ \times 9 \\ \hline \end{array}$$

$$\begin{array}{r} 91 \\ \times 3 \\ \hline \end{array} \qquad \begin{array}{r} 69 \\ \times 3 \\ \hline \end{array} \qquad \begin{array}{r} 63 \\ \times 9 \\ \hline \end{array}$$

$$\begin{array}{r} 83 \\ \times 9 \\ \hline \end{array} \qquad \begin{array}{r} 68 \\ \times 6 \\ \hline \end{array} \qquad \begin{array}{r} 55 \\ \times 9 \\ \hline \end{array}$$

$$\begin{array}{r} 83 \\ \times 3 \\ \hline \end{array} \qquad \begin{array}{r} 41 \\ \times 3 \\ \hline \end{array} \qquad \begin{array}{r} 84 \\ \times 1 \\ \hline \end{array}$$

mathbasics (1,2,3 ... to Pre-Algebra) Target time: 2 min Student's time: ____ min

Solve the following problems:

$$\begin{array}{r} 40 \\ \times 8 \\ \hline \end{array} \qquad \begin{array}{r} 48 \\ \times 6 \\ \hline \end{array} \qquad \begin{array}{r} 50 \\ \times 3 \\ \hline \end{array}$$

$$\begin{array}{r} 75 \\ \times 4 \\ \hline \end{array} \qquad \begin{array}{r} 14 \\ \times 8 \\ \hline \end{array} \qquad \begin{array}{r} 42 \\ \times 3 \\ \hline \end{array}$$

$$\begin{array}{r} 83 \\ \times 3 \\ \hline \end{array} \qquad \begin{array}{r} 65 \\ \times 5 \\ \hline \end{array} \qquad \begin{array}{r} 63 \\ \times 2 \\ \hline \end{array}$$

$$\begin{array}{r} 50 \\ \times 4 \\ \hline \end{array} \qquad \begin{array}{r} 27 \\ \times 1 \\ \hline \end{array} \qquad \begin{array}{r} 16 \\ \times 7 \\ \hline \end{array}$$

$$\begin{array}{r} 14 \\ \times 8 \\ \hline \end{array} \qquad \begin{array}{r} 98 \\ \times 4 \\ \hline \end{array} \qquad \begin{array}{r} 28 \\ \times 8 \\ \hline \end{array}$$

Multiplication, Level 4

Unit: 15
Worksheet: 1

mathbasics (1,2,3 ... to Pre-Algebra)

Target time: 4 min **Student's time: ____ min**

Solve the following problems:

$$\begin{array}{r} 87 \\ \times 33 \\ \hline \end{array} \qquad \begin{array}{r} 77 \\ \times 97 \\ \hline \end{array} \qquad \begin{array}{r} 87 \\ \times 13 \\ \hline \end{array}$$

$$\begin{array}{r} 75 \\ \times 97 \\ \hline \end{array} \qquad \begin{array}{r} 19 \\ \times 79 \\ \hline \end{array} \qquad \begin{array}{r} 10 \\ \times 16 \\ \hline \end{array}$$

$$\begin{array}{r} 94 \\ \times 66 \\ \hline \end{array} \qquad \begin{array}{r} 24 \\ \times 65 \\ \hline \end{array} \qquad \begin{array}{r} 26 \\ \times 46 \\ \hline \end{array}$$

$$\begin{array}{r} 54 \\ \times 40 \\ \hline \end{array} \qquad \begin{array}{r} 13 \\ \times 14 \\ \hline \end{array} \qquad \begin{array}{r} 25 \\ \times 37 \\ \hline \end{array}$$

$$\begin{array}{r} 64 \\ \times 56 \\ \hline \end{array} \qquad \begin{array}{r} 59 \\ \times 41 \\ \hline \end{array} \qquad \begin{array}{r} 56 \\ \times 66 \\ \hline \end{array}$$

Multiplication, Level 4

Unit: 15
Worksheet: 2

mathbasics (1,2,3 ... to Pre-Algebra) **Target time: 4 min** **Student's time: ____ min**

Solve the following problems:

40×43	89×18	41×77
21×16	74×30	86×74
36×80	40×51	46×56
88×91	87×91	95×13
29×60	59×78	91×15

Multiplication, Level 4

Unit: 15
Worksheet: 3

mathbasics (1,2,3 ... to Pre-Algebra) **Target time: 4 min** **Student's time: ____ min**

Solve the following problems:

$\begin{array}{r} 44 \\ \times 32 \\ \hline \end{array}$ $\begin{array}{r} 49 \\ \times 34 \\ \hline \end{array}$ $\begin{array}{r} 40 \\ \times 80 \\ \hline \end{array}$

$\begin{array}{r} 11 \\ \times 51 \\ \hline \end{array}$ $\begin{array}{r} 87 \\ \times 75 \\ \hline \end{array}$ $\begin{array}{r} 72 \\ \times 73 \\ \hline \end{array}$

$\begin{array}{r} 50 \\ \times 99 \\ \hline \end{array}$ $\begin{array}{r} 54 \\ \times 81 \\ \hline \end{array}$ $\begin{array}{r} 51 \\ \times 90 \\ \hline \end{array}$

$\begin{array}{r} 38 \\ \times 40 \\ \hline \end{array}$ $\begin{array}{r} 81 \\ \times 26 \\ \hline \end{array}$ $\begin{array}{r} 31 \\ \times 77 \\ \hline \end{array}$

$\begin{array}{r} 29 \\ \times 50 \\ \hline \end{array}$ $\begin{array}{r} 38 \\ \times 78 \\ \hline \end{array}$ $\begin{array}{r} 29 \\ \times 29 \\ \hline \end{array}$

Target time: 6 min Student's time: ____ min

Solve the following problems:

$$\begin{array}{r} 400 \\ \times\ 61 \\ \hline \end{array} \qquad \begin{array}{r} 991 \\ \times\ 74 \\ \hline \end{array} \qquad \begin{array}{r} 156 \\ \times\ 75 \\ \hline \end{array}$$

$$\begin{array}{r} 905 \\ \times\ 22 \\ \hline \end{array} \qquad \begin{array}{r} 797 \\ \times\ 60 \\ \hline \end{array} \qquad \begin{array}{r} 864 \\ \times\ 93 \\ \hline \end{array}$$

$$\begin{array}{r} 124 \\ \times\ 83 \\ \hline \end{array} \qquad \begin{array}{r} 638 \\ \times\ 50 \\ \hline \end{array} \qquad \begin{array}{r} 838 \\ \times\ 78 \\ \hline \end{array}$$

$$\begin{array}{r} 747 \\ \times\ 96 \\ \hline \end{array} \qquad \begin{array}{r} 737 \\ \times\ 57 \\ \hline \end{array} \qquad \begin{array}{r} 262 \\ \times\ 89 \\ \hline \end{array}$$

$$\begin{array}{r} 622 \\ \times\ 63 \\ \hline \end{array} \qquad \begin{array}{r} 510 \\ \times\ 33 \\ \hline \end{array} \qquad \begin{array}{r} 316 \\ \times\ 99 \\ \hline \end{array}$$

Multiplication, Level 5

Unit: 16
Worksheet: 2

mathbasics (1,2,3 ... to Pre-Algebra) **Target time: 6 min Student's time: ____ min**

Solve the following problems:

778 × 61	615 × 76	360 × 67
523 × 26	796 × 32	769 × 66
259 × 79	494 × 79	295 × 33
582 × 94	297 × 31	517 × 45
216 × 13	990 × 62	376 × 30

Multiplication, Level 5

Unit: 16
Worksheet: 3

mathbasics (1,2,3 ... to Pre-Algebra) **Target time: 6 min** **Student's time: ____ min**

Solve the following problems:

$\begin{array}{r} 623 \\ \times 15 \\ \hline \end{array}$	$\begin{array}{r} 823 \\ \times 23 \\ \hline \end{array}$	$\begin{array}{r} 824 \\ \times 18 \\ \hline \end{array}$
$\begin{array}{r} 811 \\ \times 34 \\ \hline \end{array}$	$\begin{array}{r} 349 \\ \times 60 \\ \hline \end{array}$	$\begin{array}{r} 569 \\ \times 11 \\ \hline \end{array}$
$\begin{array}{r} 269 \\ \times 72 \\ \hline \end{array}$	$\begin{array}{r} 813 \\ \times 66 \\ \hline \end{array}$	$\begin{array}{r} 527 \\ \times 10 \\ \hline \end{array}$
$\begin{array}{r} 896 \\ \times 10 \\ \hline \end{array}$	$\begin{array}{r} 951 \\ \times 19 \\ \hline \end{array}$	$\begin{array}{r} 328 \\ \times 46 \\ \hline \end{array}$
$\begin{array}{r} 553 \\ \times 44 \\ \hline \end{array}$	$\begin{array}{r} 507 \\ \times 54 \\ \hline \end{array}$	$\begin{array}{r} 971 \\ \times 78 \\ \hline \end{array}$

Multiplication, Level 6

Unit: 17
Worksheet: 1

mathbasics (1,2,3 ... to Pre-Algebra) **Target time: 12 min** **Student's time: ____ min**

Solve the following problems:

$$\begin{array}{r} 1617 \\ \times 832 \\ \hline \end{array}$$

$$\begin{array}{r} 9298 \\ \times 210 \\ \hline \end{array}$$

$$\begin{array}{r} 2606 \\ \times 599 \\ \hline \end{array}$$

$$\begin{array}{r} 6459 \\ \times 604 \\ \hline \end{array}$$

$$\begin{array}{r} 6003 \\ \times 872 \\ \hline \end{array}$$

$$\begin{array}{r} 9236 \\ \times 495 \\ \hline \end{array}$$

$$\begin{array}{r} 7587 \\ \times 658 \\ \hline \end{array}$$

$$\begin{array}{r} 4905 \\ \times 679 \\ \hline \end{array}$$

$$\begin{array}{r} 9244 \\ \times 291 \\ \hline \end{array}$$

$$\begin{array}{r} 1990 \\ \times 644 \\ \hline \end{array}$$

$$\begin{array}{r} 3797 \\ \times 719 \\ \hline \end{array}$$

$$\begin{array}{r} 7511 \\ \times 512 \\ \hline \end{array}$$

Multiplication, Level 6

Unit: 17
Worksheet: 2

mathbasics (1,2,3 ... to Pre-Algebra) **Target time: 12 min** **Student's time: ____ min**

Solve the following problems:

$\begin{array}{r} 9347 \\ \times\ 165 \\ \hline \end{array}$	$\begin{array}{r} 2170 \\ \times\ 856 \\ \hline \end{array}$	$\begin{array}{r} 3364 \\ \times\ 554 \\ \hline \end{array}$
$\begin{array}{r} 4735 \\ \times\ 398 \\ \hline \end{array}$	$\begin{array}{r} 3875 \\ \times\ 403 \\ \hline \end{array}$	$\begin{array}{r} 5085 \\ \times\ 548 \\ \hline \end{array}$
$\begin{array}{r} 9031 \\ \times\ 154 \\ \hline \end{array}$	$\begin{array}{r} 1525 \\ \times\ 503 \\ \hline \end{array}$	$\begin{array}{r} 9271 \\ \times\ 976 \\ \hline \end{array}$
$\begin{array}{r} 8988 \\ \times\ 685 \\ \hline \end{array}$	$\begin{array}{r} 6345 \\ \times\ 389 \\ \hline \end{array}$	$\begin{array}{r} 3650 \\ \times\ 559 \\ \hline \end{array}$

Multiplication, Level 6

Unit: 17
Worksheet: 3

mathbasics (1,2,3 ... to Pre-Algebra) **Target time: 12 min** **Student's time: ____ min**

Solve the following problems:

5807×464	2035×860	1837×854
3729×118	9200×490	8749×256
9449×348	5546×332	6518×963
7807×554	1175×833	9585×944

Division, Level 1

Unit: 18
Worksheet: 1

mathbasics (1,2,3 ... to Pre-Algebra) **Target time: 2 min** **Student's time: ____ min**

Solve the following problems:

$8\overline{)5}$	$7\overline{)2}$	$4\overline{)6}$
$3\overline{)7}$	$3\overline{)7}$	$1\overline{)8}$
$2\overline{)8}$	$2\overline{)3}$	$5\overline{)6}$
$2\overline{)3}$	$2\overline{)3}$	$2\overline{)4}$
$5\overline{)6}$	$2\overline{)4}$	$3\overline{)8}$

Division, Level 1

Unit: 18
Worksheet: 2

mathbasics (1,2,3 ... to Pre-Algebra) **Target time: 2 min** **Student's time: ____ min**

Solve the following problems:

$1\overline{)5}$	$1\overline{)2}$	$2\overline{)8}$
$5\overline{)7}$	$2\overline{)1}$	$1\overline{)7}$
$1\overline{)2}$	$6\overline{)7}$	$1\overline{)2}$
$1\overline{)8}$	$1\overline{)2}$	$2\overline{)4}$
$8\overline{)3}$	$2\overline{)9}$	$9\overline{)2}$

Division, Level 1

Unit: 18
Worksheet: 3

mathbasics (1,2,3 ... to Pre-Algebra) **Target time: 2 min** **Student's time: ____ min**

Solve the following problems:

$8\overline{)9}$	$4\overline{)5}$	$3\overline{)4}$
$2\overline{)3}$	$4\overline{)7}$	$4\overline{)6}$
$5\overline{)6}$	$4\overline{)6}$	$5\overline{)8}$
$6\overline{)9}$	$4\overline{)8}$	$8\overline{)9}$
$2\overline{)5}$	$9\overline{)5}$	$2\overline{)7}$

Division, Level 2

Unit: 19
Worksheet: 1

mathbasics (1,2,3 ... to Pre-Algebra)   **Target time: 2 min**   **Student's time: ____ min**

Solve the following problems:

$4\overline{)923}$	$7\overline{)786}$	$2\overline{)918}$
$2\overline{)910}$	$8\overline{)892}$	$5\overline{)852}$
$4\overline{)538}$	$1\overline{)991}$	$5\overline{)503}$

Division, Level 2

Unit: 19
Worksheet: 2

mathbasics (1,2,3 ... to Pre-Algebra) **Target time: 2 min** **Student's time: ____ min**

Solve the following problems:

$3\overline{)934}$	$1\overline{)964}$	$8\overline{)916}$
$2\overline{)389}$	$3\overline{)342}$	$2\overline{)829}$
$5\overline{)698}$	$3\overline{)397}$	$2\overline{)545}$

Division, Level 2

Unit: 19
Worksheet: 3

mathbasics (1,2,3 ... to Pre-Algebra) **Target time: 2 min** **Student's time: ____ min**

Solve the following problems:

$4\overline{)582}$	$5\overline{)840}$	$2\overline{)240}$
$5\overline{)913}$	$5\overline{)649}$	$2\overline{)729}$
$7\overline{)779}$	$4\overline{)460}$	$7\overline{)919}$

Solve the following problems:

$36\overline{)8042}$	$83\overline{)8455}$	$57\overline{)5670}$
$31\overline{)3143}$	$57\overline{)3890}$	$76\overline{)3372}$
$66\overline{)6536}$	$43\overline{)7218}$	$12\overline{)2269}$

Division, Level 3

Unit: 20
Worksheet: 2

mathbasics (1,2,3 ... to Pre-Algebra) **Target time: 12 min Student's time: ____ min**

Solve the following problems:

$95\overline{)5465}$ $24\overline{)7595}$ $47\overline{)3314}$

$95\overline{)9959}$ $17\overline{)2467}$ $64\overline{)8470}$

$53\overline{)9098}$ $65\overline{)3691}$ $75\overline{)2307}$

Division, Level 3

Unit: 20
Worksheet: 3

mathbasics (1,2,3 ... to Pre-Algebra) **Target time: 12 min Student's time: ____ min**

Solve the following problems:

$83\overline{)9658}$	$44\overline{)4118}$	$35\overline{)2113}$
$64\overline{)9204}$	$76\overline{)9836}$	$64\overline{)7885}$
$21\overline{)6005}$	$33\overline{)3589}$	$36\overline{)7123}$

Average, Median, Mode

Unit: 21
Worksheet: 1

mathbasics (1,2,3 ... to Pre-Algebra) **Target time: 5 min** **Student's time: ____ min**

I. Calculate the Average (Mean) of the numbers:

80, 60, 80, 16

104, 58, 94, 72, 88, 70

63, 16, 28, 75, 43

II. Calculate the Median of the numbers:

53, 15, 25, 37, 63, 15

86, 71, 24, 1, 33, 80

37, 93, 9, 47, 69, 4

III. Calculate the Mode of the numbers:

57, 73, 29, 29

80, 92, 58, 58

44, 73, 9, 9, 80

Average, Median, Mode

Unit: 21
Worksheet: 2

mathbasics (1,2,3 ... to Pre-Algebra) **Target time: 5 min Student's time: ____ min**

I. Calculate the Average (Mean) of the numbers:

101, 11, 6, 80, 22

13, 46, 33, 40

88, 84, 39, 93

II. Calculate the Median of the numbers:

20, 57, 20, 75, 56

95, 7, 10, 68, 31, 15

34, 16, 21, 37, 10, 21

III. Calculate the Mode of the numbers:

93, 79, 45, 72, 45

88, 8, 15, 93, 31

90, 75, 12, 11, 76, 38

I. Calculate the Average (Mean) of the numbers:

56, 63, 40, 57

53, 59, 49, 24, 71, 86

66, 96, 14, 5, 66, 35

II. Calculate the Median of the numbers:

92, 12, 89, 71

71, 5, 99, 23, 98, 63

10, 17, 72, 75, 71, 14

III. Calculate the Mode of the numbers:

69, 69, 33, 43

2, 9, 45, 56, 93

92, 14, 70, 62

mathbasics (1,2,3 ... to Pre-Algebra) Target time: 5 min Student's time: ____ min

I.

4^1 7^1 4^5

II.

What power of 8 is equal to 262144?

What power of 7 is equal to 16807?

III. Write using exponents:

9 to the first power:

7 cubed:

4 to the first power:

2 to the sixth power:

2 to the first power:

9 to the sixth power:

IV.

$\sqrt[5]{243}$ $\sqrt[6]{64}$ $\sqrt[4]{16}$ $\sqrt[3]{8}$

Powers, Roots

Unit: 22
Worksheet: 2

mathbasics (1,2,3 ... to Pre-Algebra) **Target time: 5 min** **Student's time: ____ min**

I.

9^{1} 6^{4} 6^{6}

II.

What power of 5 is equal to 25?

What power of 7 is equal to 1?

III. Write using exponents:

3 to the fifth power:

4 to the first power:

6 to the sixth power:

8 to the fourth power:

3 to the power of zero:

8 to the first power:

IV.

$\sqrt[6]{64}$ $\sqrt[1]{9}$ $\sqrt[1]{9}$ $\sqrt[6]{117649}$

mathbasics (1,2,3 ... to Pre-Algebra) **Target time: 5 min Student's time: ____ min**

I.

4^4 3^6 4^5

II.

What power of 8 is equal to 32768?

What power of 3 is equal to 27?

III. Write using exponents:

8 squared:

6 to the fifth power:

5 to the first power:

3 to the power of zero:

2 squared:

6 to the first power:

IV.

$\sqrt[4]{256}$ $\sqrt{9}$ $\sqrt[1]{7}$ $\sqrt[3]{64}$

Fractions, Simple to Mixed

Unit: 23
Worksheet: 1

mathbasics (1,2,3 ... to Pre-Algebra) Target time: 12 min Student's time: ____ min

Convert the fraction to a mixed number:

$\frac{684}{12}$ =

$\frac{589}{83}$ =

$\frac{159}{52}$ =

$\frac{614}{38}$ =

$\frac{579}{15}$ =

$\frac{182}{66}$ =

$\frac{352}{21}$ =

$\frac{430}{67}$ =

$\frac{443}{81}$ =

$\frac{762}{53}$ =

$\frac{57}{44}$ =

$\frac{594}{67}$ =

Fractions, Simple to Mixed

Unit: 23
Worksheet: 2

mathbasics (1,2,3 ... to Pre-Algebra) **Target time: 12 min** **Student's time: ____ min**

Convert the fraction to a mixed number:

$\frac{496}{34} =$ $\frac{358}{29} =$ $\frac{340}{73} =$

$\frac{781}{63} =$ $\frac{790}{12} =$ $\frac{417}{26} =$

$\frac{360}{27} =$ $\frac{372}{59} =$ $\frac{710}{47} =$

$\frac{358}{76} =$ $\frac{842}{20} =$ $\frac{932}{83} =$

Fractions, Simple to Mixed

Unit: 23
Worksheet: 3

mathbasics (1,2,3 ... to Pre-Algebra) **Target time: 12 min** **Student's time: ____ min**

Convert the fraction to a mixed number:

$\frac{461}{26} =$	$\frac{376}{77} =$	$\frac{445}{64} =$
$\frac{574}{78} =$	$\frac{308}{19} =$	$\frac{794}{100} =$
$\frac{935}{39} =$	$\frac{247}{47} =$	$\frac{663}{46} =$
$\frac{963}{62} =$	$\frac{890}{85} =$	$\frac{568}{71} =$

Fractions: Mixed to Simple

Unit: 24
Worksheet: 1

Target time: 6 min Student's time: ____ min

Convert the mixed number to a simple fraction:

$3\frac{27}{607} =$	$8\frac{42}{690} =$	$20\frac{98}{632} =$
$15\frac{3}{863} =$	$8\frac{71}{488} =$	$16\frac{36}{366} =$
$7\frac{60}{586} =$	$1\frac{7}{326} =$	$12\frac{35}{243} =$
$10\frac{29}{978} =$	$11\frac{61}{994} =$	$9\frac{5}{572} =$

Fractions, Mixed to Simple

Unit: 24
Worksheet: 2

mathbasics (1,2,3 ... to Pre-Algebra) **Target time: 6 min** **Student's time: ____ min**

Convert the mixed number to a simple fraction:

$13\frac{31}{941} =$

$17\frac{82}{732} =$

$2\frac{71}{503} =$

$14\frac{85}{693} =$

$16\frac{37}{779} =$

$17\frac{8}{468} =$

$19\frac{89}{222} =$

$20\frac{15}{161} =$

$13\frac{6}{571} =$

$14\frac{63}{300} =$

$15\frac{17}{781} =$

$19\frac{96}{734} =$

Fractions, Mixed to Simple

Unit: 24
Worksheet: 3

mathbasics (1,2,3 ... to Pre-Algebra) **Target time: 6 min** **Student's time: ____ min**

Convert the mixed number to a simple fraction:

$6\frac{82}{122} =$ $8\frac{34}{460} =$ $20\frac{17}{783} =$

$19\frac{40}{504} =$ $3\frac{35}{148} =$ $18\frac{82}{777} =$

$10\frac{97}{129} =$ $14\frac{9}{645} =$ $2\frac{32}{314} =$

$15\frac{18}{562} =$ $12\frac{70}{149} =$ $1\frac{49}{765} =$

Prime Numbers, Prime Factors

Unit: 25
Worksheet: 1

mathbasics (1,2,3 ... to Pre-Algebra)

Target time: 19 min **Student's time: ____ min**

Find the prime factors, arrange in order, and express as exponent:

60 =	108 =	5670 =
214375 =	70 =	280 =
735 =	11025 =	17640 =
2625 =	1050 =	34300 =

Prime Numbers, Prime Factors

Unit: 25
Worksheet: 2

mathbasics (1,2,3 ... to Pre-Algebra) **Target time: 19 min** **Student's time: ____ min**

Find the prime factors, arrange in order, and express as exponent:

7000 =	7000 =	5145 =
441 =	50 =	36750 =
1890 =	12600 =	315 =
30 =	60 =	2835 =

Prime Numbers, Prime Factors

mathbasics (1,2,3 ... to Pre-Algebra)

Target time: 19 min **Student's time: ____ min**

Find the prime factors, arrange in order, and express as exponent:

7350 =

150 =

1323 =

23814 =

120 =

90 =

30 =

5250 =

735 =

3780 =

3780 =

138915 =

Find the GCD:

240 and 882	630 and 21875	56250 and 20580
4725 and 6174	6300 and 51450	60 and 154350
300 and 100	8 and 1125	1715 and 48020
34300 and 315	1890 and 18	18 and 2058
350 and 6300	13230 and 12348	24010 and 73500
392 and 92610	20580 and 13230	216090 and 588

Greatest Common Divisor

mathbasics (1,2,3 ... to Pre-Algebra) Target time: 36 min Student's time: ____ min

Find the GCD:

90 and 46305	1715 and 490	3600 and 51450
7350 and 28	210 and 18900	26250 and 73500
17640 and 105	3528 and 18	29400 and 16
125 and 14700	1960 and 882	68600 and 44100
63 and 750	280 and 490	4375 and 630
735 and 72030	375 and 49000	675 and 17640

Greatest Common Divisor

Unit: 26
Worksheet: 3

mathbasics (1,2,3 ... to Pre-Algebra) **Target time: 36 min** **Student's time: ____ min**

Find the GCD:

4050 and 257250	36750 and 180	150 and 36
4900 and 375	90 and 90	10290 and 14700
1372 and 20	1050 and 75	28 and 420
120 and 125	504 and 70	189 and 1029
2100 and 61740	168 and 1620	882 and 4725
1176 and 50	4200 and 1176490	245 and 61740

Reducing Fractions

Unit: 27
Worksheet: 1

mathbasics (1,2,3 ... to Pre-Algebra) **Target time: 13 min** **Student's time: ____ min**

Reduce each fraction to its lowest terms:

$\frac{75}{375} =$ $\frac{50}{200} =$ $\frac{210}{350} =$

$\frac{343}{686} =$ $\frac{70}{140} =$ $\frac{30}{240} =$

$\frac{7}{8}$ in terms of 16ths $\frac{3}{5}$ in terms of 15ths $\frac{1}{2}$ in terms of 22nds

$\frac{5}{8}$ in terms of 104ths $\frac{5}{8}$ in terms of 88ths $\frac{7}{8}$ in terms of 40ths

Reducing Fractions

Unit: 27
Worksheet: 2

mathbasics (1,2,3 ... to Pre-Algebra) **Target time: 13 min** **Student's time: ____ min**

Reduce each fraction to its lowest terms:

$\frac{105}{210} =$ $\frac{36}{96} =$ $\frac{350}{560} =$

$\frac{315}{525} =$ $\frac{28}{140} =$ $\frac{112}{140} =$

$\frac{5}{8}$ in terms of 56ths $\frac{3}{8}$ in terms of 40ths $\frac{7}{8}$ in terms of 16ths

$\frac{3}{5}$ in terms of 25ths $\frac{3}{5}$ in terms of 35ths $\frac{1}{2}$ in terms of 26ths

Reducing Fractions

Unit: 27
Worksheet: 3

mathbasics (1,2,3 ... to Pre-Algebra) **Target time: 13 min** **Student's time: ____ min**

Reduce each fraction to its lowest terms:

$\frac{225}{600} =$

$\frac{98}{196} =$

$\frac{40}{100} =$

$\frac{125}{500} =$

$\frac{90}{144} =$

$\frac{81}{108} =$

$\frac{5}{8}$ in terms of 104ths

$\frac{1}{8}$ in terms of 104ths

$\frac{2}{5}$ in terms of 35ths

$\frac{5}{8}$ in terms of 104ths

$\frac{5}{8}$ in terms of 16ths

$\frac{1}{5}$ in terms of 65ths

Fractions, Addition

Unit: 28
Worksheet: 1

mathbasics (1,2,3 ... to Pre-Algebra) **Target time: 11 min** **Student's time: ____ min**

Solve the following problems (leave the answer in reduced, improper fraction form):

$\frac{18}{87} + \frac{23}{87} =$

$\frac{19}{79} + \frac{37}{79} =$

$\frac{4}{83} + \frac{39}{83} =$

$\frac{2}{3} + \frac{16}{2} =$

$\frac{38}{3} + \frac{31}{5} =$

$\frac{12}{5} + \frac{23}{3} =$

$\frac{10}{11} + \frac{19}{3} =$

$\frac{2}{3} + \frac{30}{5} =$

$\frac{8}{11} + \frac{39}{2} =$

$8\frac{27}{3} + 7\frac{19}{11} =$

$2\frac{21}{7} + 7\frac{19}{11} =$

$2\frac{39}{11} + 8\frac{17}{3} =$

Fractions, Addition

Unit: 28
Worksheet: 2

mathbasics (1,2,3 ... to Pre-Algebra) **Target time: 11 min** **Student's time: ____ min**

Solve the following problems (leave the answer in reduced, improper fraction form):

$\frac{22}{78}+\frac{39}{78}=$

$\frac{26}{66}+\frac{35}{66}=$

$\frac{34}{77}+\frac{20}{77}=$

$\frac{20}{11}+\frac{38}{3}=$

$\frac{14}{11}+\frac{5}{5}=$

$\frac{35}{11}+\frac{21}{5}=$

$\frac{24}{7}+\frac{1}{5}=$

$\frac{1}{5}+\frac{4}{7}=$

$\frac{23}{5}+\frac{28}{2}=$

$1\frac{25}{7}+4\frac{29}{5}=$

$4\frac{2}{7}+7\frac{8}{11}=$

$5\frac{39}{3}+8\frac{16}{5}=$

Fractions, Addition

Unit: 28
Worksheet: 3

mathbasics (1,2,3 ... to Pre-Algebra) **Target time: 11 min** **Student's time: ____ min**

Solve the following problems (leave the answer in reduced, improper fraction form):

$\frac{22}{80} + \frac{1}{80} =$

$\frac{24}{91} + \frac{31}{91} =$

$\frac{23}{79} + \frac{29}{79} =$

$\frac{9}{5} + \frac{22}{7} =$

$\frac{25}{2} + \frac{19}{5} =$

$\frac{18}{11} + \frac{7}{7} =$

$\frac{35}{3} + \frac{20}{7} =$

$\frac{39}{3} + \frac{30}{7} =$

$\frac{13}{3} + \frac{25}{7} =$

$1\frac{36}{5} + 4\frac{37}{3} =$

$4\frac{36}{2} + 6\frac{16}{3} =$

$2\frac{29}{7} + 9\frac{19}{2} =$

Fractions, Subtraction

Unit: 29
Worksheet: 1

mathbasics (1,2,3 ... to Pre-Algebra) **Target time: 10 min Student's time: ____ min**

Solve the following problems (leave the answer in reduced, improper fraction form):

$\frac{36}{83} - \frac{13}{83} =$

$\frac{33}{76} - \frac{12}{76} =$

$\frac{16}{51} - \frac{9}{51} =$

$\frac{11}{6} - \frac{17}{15} =$

$\frac{15}{5} - \frac{9}{6} =$

$\frac{4}{4} - \frac{8}{14} =$

$\frac{7}{6} - \frac{3}{3} =$

$\frac{4}{2} - \frac{9}{13} =$

$\frac{13}{3} - \frac{15}{6} =$

$8\frac{4}{2} - 2\frac{10}{8} =$

$9\frac{1}{9} - 1\frac{6}{3} =$

$6\frac{5}{5} - 2\frac{6}{2} =$

Fractions, Subtraction

Unit: 29
Worksheet: 2

mathbasics (1,2,3 ... to Pre-Algebra) **Target time: 10 min Student's time: ____ min**

Solve the following problems (leave the answer in reduced, improper fraction form):

$\frac{28}{93} - \frac{17}{93} =$

$\frac{19}{51} - \frac{16}{51} =$

$\frac{39}{80} - \frac{1}{80} =$

$\frac{12}{3} - \frac{13}{13} =$

$\frac{15}{6} - \frac{1}{14} =$

$\frac{12}{4} - \frac{2}{4} =$

$\frac{9}{2} - \frac{19}{7} =$

$\frac{13}{6} - \frac{5}{12} =$

$\frac{7}{5} - \frac{10}{10} =$

$11\frac{10}{3} - 2\frac{11}{4} =$

$12\frac{6}{8} - 3\frac{4}{9} =$

$11\frac{5}{7} - 4\frac{5}{5} =$

Fractions, Subtraction

Unit: 29
Worksheet: 3

mathbasics (1,2,3 ... to Pre-Algebra)   **Target time: 10 min   Student's time: ____ min**

Solve the following problems (leave the answer in reduced, improper fraction form):

$\frac{33}{88} - \frac{4}{88} =$ $\frac{35}{72} - \frac{8}{72} =$ $\frac{23}{78} - \frac{5}{78} =$

$\frac{9}{3} - \frac{11}{11} =$ $\frac{17}{5} - \frac{6}{3} =$ $\frac{4}{5} - \frac{2}{12} =$

$\frac{2}{3} - \frac{5}{9} =$ $\frac{13}{6} - \frac{1}{4} =$ $\frac{16}{3} - \frac{13}{12} =$

$15\frac{2}{10} - 4\frac{10}{11} =$ $10\frac{4}{3} - 3\frac{10}{9} =$ $10\frac{10}{10} - 3\frac{3}{8} =$

Least Common Denominator

Unit: 30
Worksheet: 1

mathbasics (1,2,3 ... to Pre-Algebra) **Target time: 25 min** **Student's time: ____ min**

Find the LCD and Solve the following problems (leave the answer in improper fraction form):

$\frac{8}{21} + \frac{7}{42} =$

LCD of 21 and 42 = ____

$\frac{3}{98} + \frac{4}{35} =$

LCD of 98 and 35 = ____

$\frac{3}{25} + \frac{7}{21} =$

LCD of 25 and 21 = ____

$\frac{6}{21} - \frac{1}{45} =$

LCD of 21 and 45 = ____

$\frac{9}{42} - \frac{7}{105} =$

LCD of 42 and 105 = ____

$\frac{3}{70} + \frac{8}{35} =$

LCD of 70 and 35 = ____

$4\frac{5}{18} - 3\frac{7}{42} =$

LCD of 18 and 42 = ____

$2\frac{8}{15} + 3\frac{7}{45} =$

LCD of 15 and 45 = ____

$1\frac{1}{49} + 1\frac{1}{12} =$

LCD of 49 and 12 = ____

$4\frac{4}{14} + 1\frac{4}{15} =$

LCD of 14 and 15 = ____

$2\frac{9}{30} + 5\frac{1}{343} =$

LCD of 30 and 343 = ____

$4\frac{4}{10} + 1\frac{1}{105} =$

LCD of 10 and 105 = ____

Least Common Denominator

Unit: 30
Worksheet: 2

mathbasics (1,2,3 ... to Pre-Algebra)

Target time: 25 min Student's time: ____ min

Find the LCD and Solve the following problems (leave the answer in improper fraction form):

$\frac{8}{21} - \frac{5}{70} =$

LCD of 21 and 70 = ____

$\frac{2}{15} - \frac{2}{70} =$

LCD of 15 and 70 = ____

$\frac{8}{18} - \frac{5}{35} =$

LCD of 18 and 35 = ____

$\frac{8}{30} + \frac{2}{35} =$

LCD of 30 and 35 = ____

$\frac{8}{21} + \frac{3}{45} =$

LCD of 21 and 45 = ____

$\frac{3}{98} + \frac{2}{20} =$

LCD of 98 and 20 = ____

$2\frac{2}{6} - 1\frac{4}{10} =$

LCD of 6 and 10 = ____

$1\frac{6}{14} + 5\frac{2}{105} =$

LCD of 14 and 105 = ____

$5\frac{3}{12} - 2\frac{1}{147} =$

LCD of 12 and 147 = ____

$2\frac{9}{10} - 2\frac{7}{30} =$

LCD of 10 and 30 = ____

$4\frac{9}{70} - 3\frac{5}{10} =$

LCD of 70 and 10 = ____

$5\frac{2}{35} - 3\frac{4}{9} =$

LCD of 35 and 9 = ____

Least Common Denominator

Unit: 30
Worksheet: 3

mathbasics (1,2,3 ... to Pre-Algebra) **Target time: 25 min Student's time: ____ min**

Find the LCD and Solve the following problems (leave the answer in improper fraction form):

$\frac{1}{98} - \frac{3}{343} =$

LCD of 98 and 343 = ____

$\frac{7}{4} - \frac{2}{21} =$

LCD of 4 and 21 = ____

$\frac{5}{21} - \frac{6}{70} =$

LCD of 21 and 70 = ____

$\frac{9}{8} + \frac{2}{35} =$

LCD of 8 and 35 = ____

$\frac{4}{21} - \frac{3}{30} =$

LCD of 21 and 30 = ____

$\frac{9}{35} - \frac{3}{28} =$

LCD of 35 and 28 = ____

$5\frac{7}{21} + 3\frac{5}{35} =$

LCD of 21 and 35 = ____

$4\frac{8}{15} - 2\frac{1}{21} =$

LCD of 15 and 21 = ____

$2\frac{7}{21} - 2\frac{3}{63} =$

LCD of 21 and 63 = ____

$3\frac{5}{15} - 1\frac{4}{10} =$

LCD of 15 and 10 = ____

$2\frac{7}{45} + 1\frac{2}{70} =$

LCD of 45 and 70 = ____

$5\frac{9}{98} - 3\frac{1}{35} =$

LCD of 98 and 35 = ____

Fractions, Multiplication

Unit: 31
Worksheet: 1

mathbasics (1,2,3 ... to Pre-Algebra) **Target time: 3 min** **Student's time: ____ min**

Multiply the fractions and reduce to simplest fraction:

$\frac{10}{10} \times \frac{5}{6} =$

$\frac{7}{8} \times \frac{8}{6} =$

$\frac{2}{7} \times \frac{3}{2} =$

$\frac{10}{3} \times \frac{8}{5} =$

$\frac{2}{4} \times \frac{7}{7} =$

$\frac{6}{2} \times \frac{7}{5} =$

$\frac{7}{5} \times \frac{5}{9} =$

$\frac{7}{3} \times \frac{1}{8} =$

$\frac{4}{7} \times \frac{2}{6} =$

$\frac{9}{10} \times \frac{7}{4} =$

$\frac{5}{10} \times \frac{7}{7} =$

$\frac{8}{4} \times \frac{7}{2} =$

Fractions, Multiplication

Unit: 31
Worksheet: 2

mathbasics (1,2,3 ... to Pre-Algebra) Target time: 3 min Student's time: ____ min

Multiply the fractions and reduce to simplest fraction:

$\frac{8}{3} \times \frac{6}{9} =$

$\frac{6}{8} \times \frac{2}{5} =$

$\frac{5}{2} \times \frac{3}{6} =$

$\frac{6}{8} \times \frac{7}{10} =$

$\frac{5}{6} \times \frac{2}{8} =$

$\frac{1}{9} \times \frac{8}{4} =$

$\frac{7}{10} \times \frac{8}{4} =$

$\frac{8}{7} \times \frac{9}{6} =$

$\frac{1}{4} \times \frac{3}{8} =$

$\frac{8}{9} \times \frac{2}{4} =$

$\frac{6}{5} \times \frac{10}{8} =$

$\frac{10}{7} \times \frac{10}{3} =$

Fractions, Multiplication

Unit: 31
Worksheet: 3

mathbasics (1,2,3 ... to Pre-Algebra) **Target time: 3 min** **Student's time: ____ min**

Multiply the fractions and reduce to simplest fraction:

$\frac{9}{9} \times \frac{4}{10} =$

$\frac{7}{9} \times \frac{8}{9} =$

$\frac{5}{9} \times \frac{10}{3} =$

$\frac{10}{9} \times \frac{1}{6} =$

$\frac{2}{5} \times \frac{6}{3} =$

$\frac{1}{9} \times \frac{6}{3} =$

$\frac{1}{3} \times \frac{7}{8} =$

$\frac{2}{6} \times \frac{7}{2} =$

$\frac{9}{8} \times \frac{4}{7} =$

$\frac{3}{9} \times \frac{1}{8} =$

$\frac{10}{3} \times \frac{4}{2} =$

$\frac{6}{8} \times \frac{7}{4} =$

Fractions, Division

Unit: 32
Worksheet: 1

mathbasics (1,2,3 ... to Pre-Algebra)   **Target time: 3 min   Student's time: ____ min**

Solve the following problems and reduce to simplest fraction:

$\frac{7}{3} \div \frac{4}{10} =$ $\frac{4}{9} \div \frac{3}{7} =$ $\frac{2}{10} \div \frac{5}{2} =$

$\frac{10}{8} \div \frac{10}{4} =$ $\frac{9}{7} \div \frac{6}{9} =$ $\frac{2}{4} \div \frac{7}{3} =$

$\frac{7}{5} \div \frac{5}{9} =$ $\frac{7}{5} \div \frac{8}{7} =$ $\frac{3}{2} \div \frac{6}{2} =$

$\frac{3}{5} \div \frac{9}{5} =$ $\frac{7}{8} \div \frac{10}{4} =$ $\frac{7}{4} \div \frac{4}{6} =$

Fractions, Division

Unit: 32
Worksheet: 2

mathbasics (1,2,3 ... to Pre-Algebra) **Target time: 3 min Student's time: ____ min**

Solve the following problems and reduce to simplest fraction:

$\frac{8}{3} \div \frac{10}{6} =$

$\frac{10}{10} \div \frac{7}{6} =$

$\frac{2}{6} \div \frac{10}{10} =$

$\frac{6}{5} \div \frac{7}{9} =$

$\frac{8}{6} \div \frac{10}{4} =$

$\frac{2}{8} \div \frac{8}{7} =$

$\frac{6}{10} \div \frac{6}{9} =$

$\frac{4}{6} \div \frac{3}{10} =$

$\frac{7}{6} \div \frac{10}{7} =$

$\frac{4}{6} \div \frac{3}{6} =$

$\frac{9}{2} \div \frac{6}{3} =$

$\frac{3}{9} \div \frac{3}{8} =$

Fractions, Division

Unit: 32
Worksheet: 3

mathbasics (1,2,3 ... to Pre-Algebra)

Target time: 3 min   Student's time: ____ min

Solve the following problems and reduce to simplest fraction:

$\frac{9}{5} \div \frac{2}{4} =$

$\frac{4}{8} \div \frac{3}{7} =$

$\frac{7}{3} \div \frac{2}{10} =$

$\frac{5}{5} \div \frac{10}{4} =$

$\frac{2}{6} \div \frac{5}{2} =$

$\frac{5}{4} \div \frac{9}{3} =$

$\frac{4}{7} \div \frac{8}{4} =$

$\frac{5}{3} \div \frac{3}{6} =$

$\frac{10}{5} \div \frac{10}{6} =$

$\frac{6}{2} \div \frac{8}{8} =$

$\frac{4}{9} \div \frac{2}{10} =$

$\frac{8}{8} \div \frac{5}{10} =$

Simplification of Fractions

Unit: 33
Worksheet: 1

mathbasics (1,2,3 ... to Pre-Algebra) **Target time: 11 min** **Student's time: ____ min**

Solve each problem and reduce to lowest simple fraction:

$\frac{12}{5} \times \frac{15}{2} \div \frac{5}{2} =$

$\frac{6}{60} \times \frac{24}{4} \times \frac{2}{25} =$

$\frac{9}{3} \div \frac{375}{30} \div \frac{100}{225} =$

$\frac{3}{5} \times \frac{6}{4} \div \frac{18}{5} =$

$\frac{3}{30} \div \frac{375}{18} \div \frac{4}{2} =$

$\frac{45}{40} \div \frac{5}{6} \times \frac{5}{45} =$

$\frac{5}{15} \div \frac{45}{30} \div \frac{30}{135} =$

$\frac{90}{9} \div \frac{60}{9} \times \frac{4}{125} =$

$\frac{3}{3} \times \frac{12}{4} \times \frac{20}{15} =$

$\frac{60}{9} \div \frac{30}{10} \div \frac{30}{3} =$

Simplification of Fractions

Unit: 33
Worksheet: 2

mathbasics (1,2,3 ... to Pre-Algebra) **Target time: 11 min** **Student's time: ____ min**

Solve each problem and reduce to lowest simple fraction:

$\frac{4}{5} \div \frac{30}{6} \div \frac{10}{15} =$

$\frac{90}{12} \times \frac{3}{2} \times \frac{3}{150} =$

$\frac{100}{90} \times \frac{25}{225} \div \frac{15}{15} =$

$\frac{30}{15} \div \frac{50}{15} \div \frac{10}{2} =$

$\frac{5}{150} \times \frac{30}{36} \times \frac{2}{8} =$

$\frac{135}{90} \times \frac{12}{15} \times \frac{12}{27} =$

$\frac{3}{5} \times \frac{3}{100} \times \frac{30}{6} =$

$\frac{75}{150} \times \frac{9}{150} \div \frac{2}{225} =$

$\frac{4}{250} \times \frac{100}{9} \times \frac{24}{18} =$

$\frac{15}{36} \times \frac{25}{12} \div \frac{12}{60} =$

Simplification of Fractions

Unit: 33
Worksheet: 3

mathbasics (1,2,3 ... to Pre-Algebra)   Target time: 11 min   Student's time: ____ min

Solve each problem and reduce to lowest simple fraction:

$\frac{90}{225} \div \frac{3}{90} \div \frac{20}{75} =$

$\frac{100}{24} \times \frac{90}{2} \div \frac{3}{45} =$

$\frac{30}{15} \div \frac{75}{15} \div \frac{5}{6} =$

$\frac{20}{250} \times \frac{3}{3} \div \frac{15}{6} =$

$\frac{24}{3} \div \frac{12}{18} \times \frac{100}{6} =$

$\frac{2}{30} \div \frac{15}{50} \times \frac{50}{3} =$

$\frac{6}{2} \div \frac{150}{60} \times \frac{6}{30} =$

$\frac{150}{90} \times \frac{24}{12} \times \frac{6}{250} =$

$\frac{5}{15} \times \frac{6}{4} \times \frac{9}{24} =$

$\frac{3}{20} \times \frac{250}{25} \times \frac{18}{27} =$

Complex Fractions

Unit: 34
Worksheet: 1

mathbasics (1,2,3 ... to Pre-Algebra) **Target time: 11 min** **Student's time: ____ min**

Solve the following problems:

$$\frac{\frac{7}{3} \div \frac{1}{2} \times 3}{\frac{1}{5} \div \frac{1}{2} \times 4} =$$

$$\frac{6\frac{1}{3} + 7\frac{2}{5} + 9\frac{1}{2}}{2\frac{1}{2} \div \frac{6}{3} \div 4} =$$

Complex Fractions

Unit: 34
Worksheet: 2

Target time: 11 min Student's time: ____ min

Solve the following problems:

$$\frac{\frac{4}{5} \div \frac{30}{6} \div \frac{10}{15}}{\frac{90}{12} \times \frac{3}{2} \times \frac{3}{150}} =$$

$$\frac{\frac{8}{21} + \frac{7}{42} + \frac{3}{98}}{\frac{6}{21} - \frac{1}{4} + \frac{9}{42}} =$$

Complex Fractions

Unit: 34
Worksheet: 3

mathbasics (1,2,3 ... to Pre-Algebra) **Target time: 11 min** **Student's time: ____ min**

Solve the following problems:

$$\frac{\frac{1}{2} + \frac{2}{3} + \frac{3}{5}}{\frac{3}{5} + \frac{2}{3} - \frac{1}{2}} =$$

$$\frac{\frac{8}{3} \div \frac{4}{6} \div \frac{9}{5}}{\frac{7}{3} + \frac{8}{2} - \frac{2}{7}} =$$

Laws of Addition

Unit: 35
Worksheet: 1

mathbasics (1,2,3 ... to Pre-Algebra) **Target time: 2 min Student's time: ____ min**

Which law of addition is illustrated below?

$47 + (14 + 7) = (47 + 14) + 7$

$95 + 27 = 27 + 95$

$67 + 69 = 69 + 67$

$45 + (44 + 26) = (45 + 44) + 26$

$64 + 82 = 82 + 64$

$85 + (25 + 92) = (85 + 25) + 92$

$(71 + 83) + 58 = 71 + (83 + 58)$

$(92 + 14) + 6 = 92 + (14 + 6)$

Use the indicated law to rewrite the expression:

Commutative: $30 + 34$

Commutative: $47 + 76$

Commutative: $22 + 76$

Associative: $(56 + 82) + 42$

Associative: $(3 + 98) + 78$

Commutative: $97 + 18$

Associative: $29 + (76 + 32)$

Associative: $(92 + 14) + 52$

Laws of Addition

mathbasics (1,2,3 ... to Pre-Algebra) **Target time: 2 min** **Student's time: ____ min**

Which law of addition is illustrated below?

$(57 + 56) + 59 = 57 + (56 + 59)$

$39 + (15 + 39) = (39 + 15) + 39$

$(40 + 96) + 38 = 40 + (96 + 38)$

$42 + 18 = 18 + 42$

$38 + 85 = 85 + 38$

$39 + 55 = 55 + 39$

$55 + 90 = 90 + 55$

$(11 + 7) + 82 = 11 + (7 + 82)$

Use the indicated law to rewrite the expression:

Associative: $32 + (28 + 10)$

Commutative: $15 + 61$

Commutative: $69 + 34$

Commutative: $80 + 66$

Commutative: $72 + 89$

Commutative: $63 + 28$

Commutative: $31 + 34$

Associative: $(40 + 60) + 47$

Laws of Addition

Unit: 35
Worksheet: 3

mathbasics (1,2,3 ... to Pre-Algebra) **Target time: 2 min** **Student's time: ____ min**

Which law of addition is illustrated below?

$67 + 56 = 56 + 67$

$(98 + 88) + 63 = 98 + (88 + 63)$

$36 + (60 + 91) = (36 + 60) + 91$

$(45 + 56) + 52 = 45 + (56 + 52)$

$(46 + 40) + 78 = 46 + (40 + 78)$

$31 + 2 = 2 + 31$

$39 + 7 = 7 + 39$

$(96 + 79) + 76 = 96 + (79 + 76)$

Use the indicated law to rewrite the expression:

Associative: $42 + (54 + 59)$

Commutative: $51 + 68$

Commutative: $92 + 69$

Associative: $(53 + 84) + 59$

Associative: $(88 + 1) + 89$

Associative: $61 + (69 + 87)$

Associative: $13 + (56 + 97)$

Commutative: $64 + 60$

Laws of Multiplication

Unit: 36
Worksheet: 1

mathbasics (1,2,3 ... to Pre-Algebra) Target time: 3 min Student's time: ____ min

Which law of multiplication is illustrated below?

$81 \times 79 = 79 \times 81$

$21 \times (23 \times 57) = (21 \times 23) \times 57$

$26 \times (92 + 98) = (26 \times 92) + (26 \times 98)$

$39 \times (5 \times 74) = (39 \times 5) \times 74$

$43 \times (80 + 99) = (43 \times 80) + (43 \times 99)$

$(82 \times 86) \times 54 = 82 \times (86 \times 54)$

$56 \times 1 = 1 \times 56$

$53 \times (15 + 7) = (53 \times 15) + (53 \times 7)$

Use the indicated law to rewrite the expression:

Associative: $92 \times (93 \times 90)$

Distributive: $57 \times (71 + 30)$

Commutative: 84×91

Distributive: $77 \times (85 + 13)$

Associative: $98 \times (74 \times 48)$

Commutative: 41×94

Distributive: $73 \times (23 + 73)$

Associative: $(19 \times 13) \times 57$

Laws of Multiplication

Unit: 36
Worksheet: 2

mathbasics (1,2,3 ... to Pre-Algebra) **Target time: 3 min Student's time: ____ min**

Which law of multiplication is illustrated below?

$5 \times 25 = 25 \times 5$

$89 \times (48 + 35) = (89 \times 48) + (89 \times 35)$

$(30 \times 29) \times 99 = 30 \times (29 \times 99)$

$76 \times 7 = 7 \times 76$

$(20 \times 13) \times 75 = 20 \times (13 \times 75)$

$58 \times 47 = 47 \times 58$

$35 \times 19 = 19 \times 35$

$31 \times 65 = 65 \times 31$

Use the indicated law to rewrite the expression:

Distributive: $44 \times (37 + 5)$

Associative: $(79 \times 32) \times 48$

Commutative: 70×5

Associative: $(18 \times 62) \times 31$

Commutative: 55×77

Distributive: $56 \times (74 + 20)$

Commutative: 70×33

Distributive: $77 \times (38 + 29)$

Laws of Multiplication

Unit: 36
Worksheet: 3

mathbasics (1,2,3 ... to Pre-Algebra) Target time: 3 min Student's time: ____ min

Which law of multiplication is illustrated below?

$(94 \times 59) \times 23 = 94 \times (59 \times 23)$

$36 \times (34 + 19) = (36 \times 34) + (36 \times 19)$

$29 \times 83 = 83 \times 29$

$30 \times (89 + 19) = (30 \times 89) + (30 \times 19)$

$88 \times 36 = 36 \times 88$

$43 \times (82 \times 9) = (43 \times 82) \times 9$

$43 \times (33 + 72) = (43 \times 33) + (43 \times 72)$

$84 \times 68 = 68 \times 84$

Use the indicated law to rewrite the expression:

Distributive: $62 \times (83 + 94)$

Commutative: 42×44

Associative: $51 \times (15 \times 71)$

Commutative: 14×29

Commutative: 61×58

Commutative: 40×97

Associative: $59 \times (65 \times 49)$

Associative: $(12 \times 81) \times 73$

Order of Operations

Unit: 37
Worksheet: 1

mathbasics (1,2,3 ... to Pre-Algebra) **Target time: 7 min Student's time: ____ min**

Solve the following problems:

$2 + 7 \times 3 =$

$(6 + 9) - 15 + 3 =$

$72 \div 3^2 - 2 =$

$27 \times 3 - 2^2 + (9 - 8) =$

$(2 + 5) \times 6^2 + 9 \div 3 \times 0 =$

$(3 \times 2) + 5 + 3^3 \times 9 =$

Order of Operations

Unit: 37
Worksheet: 2

mathbasics (1,2,3 ... to Pre-Algebra) **Target time: 7 min** **Student's time: ____ min**

Solve the following problems:

$5 + 7 \times 3 =$

$9 \times 9 + 6 - 2 =$

$6^2 + 7 \times 2 \times 9 - 5^3 =$

$6 + (7 - 2) + 8^2 - 35 =$

$9 \times 9 - 64 + 3^2 \div 3 =$

$7 + 2^{12} \times 2 - 5600 =$

Order of Operations

Unit: 37
Worksheet: 3

mathbasics (1,2,3 ... to Pre-Algebra) **Target time: 7 min** **Student's time: ____ min**

Solve the following problems:

$630 - 7^2 \times (3 + 4) =$

$5 \times 10 + 7^2 \times (15 \div 5) =$

$3 \times 7 + 2^5 + 9 - 55 =$

$3^3 \times (9 + 7^2 - 2) - 20 \times 0 =$

$(6 + 2) \times 4 + 10^2 - 3 =$

$3 \times (7 + 2^2) - 9 \times 2 =$

Write the following decimals in words:

0.45

0.00002

0.000013

0.00018

0.000008

0.09

Write out each of the following as decimal numbers:

fifty-one hundred-thousandths

four ten-thousandths

forty-five ten-thousandths

seventy-nine ten-thousandths

eighteen hundred-thousandths

Decimals - Introduction

Unit: 38
Worksheet: 2

mathbasics (1,2,3 ... to Pre-Algebra) **Target time: 3 min Student's time: ____ min**

Write the following decimals in words:

0.018

0.0005

0.00034

0.002

0.00005

0.000058

Write out each of the following as decimal numbers:

twenty-six hundredths

six hundred-thousandths

one ten-thousandths

ninety-seven hundred-thousandths

two hundred-thousandths

Decimals - Introduction

Unit: 38
Worksheet: 3

mathbasics (1,2,3 ... to Pre-Algebra) **Target time: 3 min** **Student's time: ____ min**

Write the following decimals in words:

0.0003

0.00004

0.048

0.07

0.0001

0.09

Write out each of the following as decimal numbers:

eight ten-thousandths

nine tenths

ninety-nine hundredths

eighty-one ten-thousandths

seventy-fiv4 millionths

Addition of Decimals

Unit: 39
Worksheet: 1

mathbasics (1,2,3 ... to Pre-Algebra)   **Target time: 11 min   Student's time: ____ min**

Solve the following problems:

$89.66 + 767.9 =$

$2.58 + 154.076023 =$

$8.9 + 2.089 =$

$952.6236 + 49.06 =$

$9.18 + 9.2 =$

$9.59 + 607.8 =$

$73.1684 + 4.5596 + 2.8237 =$

$14.5996 + 5.05061 + 19.1 =$

$65.11 + 561 + 92.406 =$

$8.2053 + 74.5 + 1.1034 =$

$4.632 + 56.834918 + 692.083 + 48.638 =$

$11.558 + 41.57 + 588.520065 + 765.54 =$

Addition of Decimals

Unit: 39
Worksheet: 2

mathbasics (1,2,3 ... to Pre-Algebra) **Target time: 11 min** **Student's time: ____ min**

Solve the following problems:

1.96 + 288.08 =

3.32 + 372.766365 =

0.22459 + 584.911 =

541.206 + 64.29

78.997805 + 9.273

76.738347 + 55.717261 =

501.0729 + 7.094 + 289.203601 =

933.9871 + 47.565 + 5.87147 =

103.31319 + 44.3 + 3 =

20.955 + 0.865 + 480.35 =

652.152 + 90.57 + 72.482097 + 471.9 =

457.8 + 482.443313 + 92.225 + 266.0394 =

Addition of Decimals

Unit: 39
Worksheet: 3

mathbasics (1,2,3 ... to Pre-Algebra) Target time: 11 min Student's time: ____ min

Solve the following problems:

67.851948 + 632.11744 =

15.5811 + 74.933 =

268.342 + 910.1571 =

641.732995 + 529.72 =

6.59 + 96.72508 =

6.381 + 85.7 =

633.404926 + 337.710264 + 75.65109 =

23.362+4.36152+98.03517=

0.32177 + 4.7 + 708.14 =

3.6 + 572.02 + 8.2507 =

58.871 + 4.582 + 290.3 + 7.36455 =

3.7 + 62.29 + 67.53583 + 467.2216 =

Subtraction of Decimals

Unit: 40
Worksheet: 1

mathbasics (1,2,3 ... to Pre-Algebra) **Target time: 12 min** **Student's time: ____ min**

Solve the following problems:

7224.50175 - 111.1 =	1689.9049 - 842.88096 =	4942.1 - 574 =
4142.1592 - 443.7866 =	1.7522 - 1.381691 =	7723.77 - 397.68916 =
149.729646 - 2.777338 =	9881.978 - 970.94 =	7979.64 - 788.21357 =
8225.69862 - 51.090485 =	5233.459973 - 835.04 =	727.1 - 7.07934 =
1806.34133 - 74.578293 =	4178.756 - 64.51405 =	7483.797063-19.59772=

Subtraction of Decimals

Unit: 40
Worksheet: 2

mathbasics (1,2,3 ... to Pre-Algebra) **Target time: 12 min Student's time: ____ min**

Solve the following problems:

5344.9 - 137.9454 =	408.57695 - 191.5 =	72.2 - 18.272632 =
9.3797 - 4.551 =	9551.0505 - 91.636764 =	5169.933932 - 26.8 =
3.740921 - 3.15274 =	673.586 - 89.79847 =	4084.3183 - 9.762 =
407.267 - 3.787881 =	8.766453 - 6.214013 =	616.057 - 24.038972 =
642.289567 - 126.559 =	623.4189 - 478.656 =	6500.011 - 862.83 =

Subtraction of Decimals

Unit: 40
Worksheet: 3

mathbasics (1,2,3 ... to Pre-Algebra) **Target time: 12 min** **Student's time: ____ min**

Solve the following problems:

506.63 - 280.4961 =	4787.37 - 29 =	5589.367 - 738.1727 =
2495.16 - 550.4 =	4.3403 - 2.976217 =	2114.958 - 56.19 =
8970.3 - 592.350273 =	7.324109 - 2.47 =	62.666191 - 32.3506 =
569.363 - 37.36 =	7294.753 - 45.6298 =	7516.196 - 583.1 =
205.6469 - 194.95 =	45.73 - 9.9437 =	8628.8106 - 915.101 =

Multiplication of Decimals

Unit: 41
Worksheet: 1

mathbasics (1,2,3 ... to Pre-Algebra) **Target time: 10 min** **Student's time: min**

Solve the following problems:

84.01 x 5.72 =	0.04 x 80.4 =	42.4 x 6.3 =
73.67 x 27.2 =	811.4 x 52.49 =	47.1 x 6.09 =
17.26 x 0.3 =	543.11 x 74.23 =	8.8 x 29.6 =
3.81 x 33.99 =	51.01 x 79.67 =	71.08 x 97.4 =

Multiplication of Decimals

Unit: 41
Worksheet: 2

mathbasics (1,2,3 ... to Pre-Algebra) Target time: 10 min Student's time: min

Solve the following problems:

644.59 x 89.8 =	19.9 x 0.9 =	6.02x61.1=
659.6 x 4.59 =	328.9 x 1 =	597.13 x 3.33 =
75.3 x 67.7 =	0 x 40.38 =	342.92 x 97.4 =
9 x 74.4 =	3.6 x 67.09 =	1.9 x 64.8 =

Multiplication of Decimals

Unit: 41
Worksheet: 3

mathbasics (1,2,3 ... to Pre-Algebra) **Target time: 10 min** **Student's time: min**

Solve the following problems:

319.8 x 6.7 =	6.9 x 56.16 =	5.4 x 17.75 =
94.19 x 24.55 =	17.64 x 4.49 =	591.9 x 8.68 =
92.2 x 53.39 =	793.68 x 32.1 =	7.6 x 3.93 =
173.75 x 32.96 =	4.2 x 25.04 =	6.81 x 2.88 =

Division of Decimals

Unit: 42
Worksheet: 1

mathbasics (1,2,3 ... to Pre-Algebra) Target time: 13 min Student's time: min

Solve the following problems:

$33 \div 2.4 =$	$105 \div 6=$	$2.1 \div 2.4=$
$0.4 \div 0.16 =$	$0.5 \div 2 =$	$5.4 \div 1.35 =$
$13.5 \div 7.5 =$	$18 \div 48 =$	$1.5 \div 1.2 =$
$63 \div 36 =$	$64.8 \div 4.05 =$	$1.04 \div 0.64 =$

Division of Decimals

Unit: 42
Worksheet: 2

mathbasics (1,2,3 ... to Pre-Algebra) Target time: 13 min Student's time: min

Solve the following problems:

$0.12 \div 0.16 =$	$2.7 \div 24 =$	$4 \div 25 =$
$31.5 \div 36 =$	$40 \div 25 =$	$0.9 \div 6 =$
$12 \div 8 =$	$0.14 \div 1.6 =$	$10 \div 2 =$
$5.4 \div 9 =$	$15 \div 12 =$	$1.26 \div 0.9 =$

Division of Decimals

Unit: 42
Worksheet: 3

mathbasics (1,2,3 ... to Pre-Algebra) Target time: 13 min Student's time: min

Solve the following problems:

$0.06 \div 0.24 =$ $3 \div 4.8 =$ $0.6 \div 3 =$

$10.5 \div 6 =$ $0.04 \div 0.16 =$ $6 \div 9.6 =$

$13 \div 80 =$ $42 \div 30 =$ $67.5 \div 60 =$

$0.4 \div 0.08 =$ $2 \div 8 =$ $2.25 \div 11.25 =$

Converting Decimals to Fractions

Unit: 43
Worksheet: 1

mathbasics (1,2,3 ... to Pre-Algebra) **Target time: 6 min** **Student's time: ____ min**

Convert the following decimals to fractions:

949.6 =	591.6 =	551.375 =
301.4 =	634.25 =	295.4 =
778.5 =	410.875 =	146.6 =
730.375 =	123.375 =	354.5 =
209.25 =	406.875 =	160.4 =

Converting Decimals to Fractions

Unit: 43
Worksheet: 2

mathbasics (1,2,3 ... to Pre-Algebra) **Target time: 6 min Student's time: ____ min**

Convert the following decimals to fractions:

306.375 =	375.875 =	283.625 =
541.625 =	141.75 =	522.75 =
276.2 =	498.75 =	938.8 =
358.125 =	752.6 =	340.8 =
479.6 =	919.625 =	33.875 =

Converting Decimals to Fractions

Unit: 43
Worksheet: 3

mathbasics (1,2,3 ... to Pre-Algebra)

Target time: 6 min Student's time: ____ min

Convert the following decimals to fractions:

953.5 =	749.125 =	69.2 =
893.125 =	304.8 =	435.875 =
999.75 =	58.8 =	936.5 =
429.75 =	730.6 =	709.625 =
470.375 =	324.125 =	609.625 =

Target time: 20 min **Student's time: ____ min**

Convert the following decimals to fractions:

$\frac{225}{125} =$ $\frac{625}{1000} =$ $\frac{36}{20} =$

$\frac{750}{1200} =$ $\frac{225}{360} =$ $\frac{125}{100} =$

$\frac{60}{48} =$ $\frac{54}{270} =$ $\frac{90}{240} =$

$\frac{36}{45} =$ $\frac{18}{144} =$ $\frac{63}{45} =$

$\frac{7.277}{1000} =$ $\frac{69.45}{10000} =$ $\frac{663.8}{10000} =$

Converting Fractions to Decimals

Unit: 44
Worksheet: 2

mathbasics (1,2,3 ... to Pre-Algebra) **Target time: 20 min Student's time: ____ min**

Convert the following decimals to fractions:

$\frac{20}{16} =$

$\frac{70}{80} =$

$\frac{450}{250} =$

$\frac{60}{96} =$

$\frac{375}{625} =$

$\frac{100}{400} =$

$\frac{60}{80} =$

$\frac{200}{125} =$

$\frac{468}{288} =$

$\frac{12}{20} =$

$\frac{132}{96} =$

$\frac{54}{135} =$

$\frac{920.4}{10} =$

$\frac{342.9}{10000} =$

$\frac{79.67}{1000} =$

Converting Fractions to Decimals

Unit: 44
Worksheet: 3

mathbasics (1,2,3 ... to Pre-Algebra) **Target time: 20 min** **Student's time: ____ min**

Convert the following decimals to fractions:

$\frac{810}{450} =$ $\frac{420}{240} =$ $\frac{132}{96} =$

$\frac{60}{96} =$ $\frac{24}{64} =$ $\frac{100}{80} =$

$\frac{54}{36} =$ $\frac{45}{25} =$ $\frac{40}{64} =$

$\frac{45}{360} =$ $\frac{702}{432} =$ $\frac{405}{540} =$

$\frac{10.19}{100} =$ $\frac{199.7}{1000} =$ $\frac{8.183}{1000} =$

Percents and Fractions

Unit: 45
Worksheet: 1

mathbasics (1,2,3 ... to Pre-Algebra) Target time: 5 min Student's time: ____ min

Convert the following percents to a fraction (reduced to lowest terms):

160% =	162.5% =	112.5% =
112.5% =	25% =	12.5% =
140% =	137.5% =	175% =

Convert the following fractions to a percent:

$\frac{63}{36}$	$\frac{28}{32}$	$\frac{6}{30}$
$\frac{24}{60}$	$\frac{264}{192}$	$\frac{14}{10}$
$\frac{12}{16}$	$\frac{54}{45}$	$\frac{3}{6}$

Percents and Fractions

Unit: 45
Worksheet: 2

mathbasics (1,2,3 ... to Pre-Algebra) **Target time: 5 min** **Student's time: ____ min**

Convert the following percents to a fraction (reduced to lowest terms):

120% = 140% = 80% =

12.5% = 180% = 180% =

25% = 112.5% = 60% =

Convert the following fractions to a percent:

$\frac{6}{24}$ $\frac{132}{96}$ $\frac{54}{90}$

$\frac{9}{12}$ $\frac{24}{60}$ $\frac{72}{60}$

$\frac{30}{48}$ $\frac{42}{24}$ $\frac{9}{36}$

Percents and Fractions

Unit: 45
Worksheet: 3

mathbasics (1,2,3 ... to Pre-Algebra) **Target time: 5 min Student's time: ____ min**

Convert the following percents to a fraction (reduced to lowest terms):

137.5% = 87.5% = 150% =

40% = 20% = 12.5% =

12.5% = 87.5% = 62.5% =

Convert the following fractions to a percent:

$\frac{81}{108}$ $\frac{324}{180}$ $\frac{162}{144}$

$\frac{54}{48}$ $\frac{54}{36}$ $\frac{14}{10}$

$\frac{21}{12}$ $\frac{6}{30}$ $\frac{6}{24}$

Percents and Decimals

Unit: 46
Worksheet: 1

mathbasics (1,2,3 ... to Pre-Algebra) Target time: 5 min Student's time: ____ min

Convert the percents to decimal:

848.2904% =	737.9254% =	4.23498% =	319.149% =
47.8% =	456.1326% =	573.36% =	155.239408% =
968.807335% =	9.9% =	824.958939% =	73.57% =
653.4% =	257.81% =	328.170513% =	202.52644% =

Convert the decimals to percent:

22.0 =	5.9 =	502.6 =	964.8440 =
0.4 =	4.6 =	9.45438 =	856.1 =
651.22 =	5.392653 =	3.1931 =	776.36 =
9.9 =	561.78441 =	1.511704 =	1.5089 =

Percents and Decimals

Unit: 46
Worksheet: 2

mathbasics (1,2,3 ... to Pre-Algebra) **Target time: 5 min** **Student's time: ____ min**

Convert the percents to decimal:

2.7629% =	62.8013% =	98.828% =	2.22% =
566.9435% =	0.3% =	686.5697% =	4.4% =
866.245% =	247.1% =	658.26176% =	551.7% =
3.5727% =	37.98260% =	1.874274% =	4.11874% =

Convert the decimals to percent:

68.951 =	55.7 =	20.229776 =	66.0 =
55.6926 =	958.63261 =	498.740 =	89.16960 =
978.5167 =	786.027661 =	24.30802 =	32.677 =
884.7759 =	9.93 =	5.829 =	874.29442 =

Percents and Decimals

Unit: 46
Worksheet: 3

mathbasics (1,2,3 ... to Pre-Algebra) **Target time: 5 min** **Student's time: ____ min**

Convert the percents to decimal:

242.5856% =	23.3% =	19.6% =	283.56564% =
42.4% =	9.95039% =	188.130% =	184.30879% =
166.2339% =	5.1114% =	195.36293% =	5.58528% =
722.46% =	1.00% =	7.28% =	27.92% =

Convert the decimals to percent:

264.4 =	2.3078 =	69.11663 =	3.464216 =
11.458372 =	30.47 =	4.4095 =	1.2 =
5.793546 =	6.293743 =	145.5523 =	605.8044 =
963.3 =	71.36 =	363.01328 =	641.91818 =

Percent Problems - Type A

Unit: 47
Worksheet: 1

mathbasics (1,2,3 ... to Pre-Algebra) **Target time: 9 min** **Student's time: min**

Solve the following problems:

77% of 58=	80% of 63 =	41% of 20 =
60% of 50=	20% of 79 =	17% of 2 =
15% of 699 =	60% of 792 =	5% of 779 =
5.16% of 9.2 =	83.4% of 7.07 =	7.64% of 6.2 =
17.3% of 2.04 =	2.8% of 1.9 =	34.1% of 74.17 =

Percent Problems - Type A

Unit: 47
Worksheet: 2

mathbasics (1,2,3 ... to Pre-Algebra) **Target time: 9 min** **Student's time: min**

Solve the following problems:

18% of 63 =	8% of 56=	8% of 25 =
50% of 68 =	84% of 95 =	76% of 7 =
36% of 588 =	71% of 346 =	83% of 166 =
78.79% of 9.5 =	71.58% of 17.6 =	97.5% of 41.3 =
4% of 27.16 =	62.9% of 8.6 =	33.8% of 8.43 =

Percent Problems - Type A

Unit: 47
Worksheet: 3

mathbasics (1,2,3 ... to Pre-Algebra) **Target time: 9 min** **Student's time: min**

Solve the following problems:

33% of 76 =	88% of 93 =	53% of 64 =
100% of 54=	1%of 58=	36% of 23 =
91% of 250 =	97% of 534 =	14% of 768 =
43.15% of 6.9 =	2.92% of 61 =	7.86% of 3.14 =
47.77% of 14.5 =	88.6% of 3.4 =	45.1% of 3.7 =

Percent Problems - Type B

Unit: 48
Worksheet: 1

mathbasics (1,2,3 ... to Pre-Algebra) **Target time: 15 min** **Student's time: ____ min**

Solve the following problems:

12 is what percent of 96?

42 is what percent of 30?

180 is what percent of 100?

4125 is what percent of 3000?

2.52 is what percent of 1.44?

5.4 is what percent of 6.75?

What percent of 72 is 99?

What percent of 72 is 90?

What percent of 600 is 825?

What percent of 180 is 252?

What percent of 9 is 2.25?

What percent of 0.6 is 1.08?

Percent Problems - Type B

Unit: 48
Worksheet: 2

mathbasics (1,2,3 ... to Pre-Algebra) **Target time: 15 min** **Student's time: ____ min**

Solve the following problems:

990 is what percent of 720?

140 is what percent of 100?

540 is what percent of 450?

25 is what percent of 20?

6.75 is what percent of 11.25?

0.04 is what percent of 0.2?

What percent of 20 is 15?

What percent of 750 is 600?

What percent of 1250 is 1750?

What percent of 1875 is 1125?

What percent of 4 is 0.5?

What percent of 1.5 is 2.25?

Percent Problems - Type B

Unit: 48
Worksheet: 3

mathbasics (1,2,3 ... to Pre-Algebra) **Target time: 15 min** **Student's time: ____ min**

Solve the following problems:

525 is what percent of 375?

2475 is what percent of 1800?

6 is what percent of 4?

240 is what percent of 150?

6.6 is what percent of 4.8?

5.4 is what percent of 4.8?

What percent of 400 is 250?

What percent of 200 is 325?

What percent of 3000 is 4875?

What percent of 144 is 162?

What percent of 0.08 is 0.14?

What percent of 7.2 is 0.9?

Percent Problems - Type C

Unit: 49
Worksheet: 1

mathbasics (1,2,3 ... to Pre-Algebra) Target time: 24 min Student's time: ____ min

Solve the following problems:

375 is 25% of what number?

252 is 87.5% of what number?

60 is 75% of what number?

36 is 20% of what number?

1.8 is 62.5% of what number?

0.21 is 87.5% of what number?

12 is 12.5% of what number?

70 is 175% of what number?

75 is 150% of what number?

12 is 60% of what number?

2.4 is 80% of what number?

6.75 is 180% of what number?

Percent Problems - Type C

Unit: 49
Worksheet: 2

mathbasics (1,2,3 ... to Pre-Algebra) **Target time: 24 min** **Student's time: ____ min**

Solve the following problems:

15 is 150% of what number?

20 is 40% of what number?

270 is 37.5% of what number?

120 is 75% of what number?

1.35 is 150% of what number?

2.4 is 160% of what number?

270 is 37.5% of what number?

216 is 180% of what number?

1080 is 160% of what number?

270 is 112.5% of what number?

0.04 is 50% of what number?

0.3 is 20% of what number?

Percent Problems - Type C

Unit: 49
Worksheet: 3

mathbasics (1,2,3 ... to Pre-Algebra) Target time: 24 min Student's time: ____ min

Solve the following problems:

120 is 125% of what number?

126 is 175% of what number?

140 is 175% of what number?

18 is 12.5% of what number?

8.1 is 120% of what number?

0.54 is 120% of what number?

90 is 50% of what number?

90 is 20% of what number?

50 is 20% of what number?

130 is 162.5% of what number?

0.1 is 125% of what number?

12.15 is 180% of what number?

Ratios

Unit: 50
Worksheet: 1

mathbasics (1,2,3 ... to Pre-Algebra) **Target time: 5 min** **Student's time: ____ min**

Solve the following problems:

1. Chris earned $45 mowing lawns. Sarah earned $50 mowing lawns. What is the ratio of Chris's money to Sarah's?

2. Jenny had 50 pencils at the beginning of the year and had 25 pencils at the end of the year. What is the ratio of pencils from the beginning of the year to the end?

3. Jimmy has 21 baseball cards and Joey has 14 baseball cards. What is the ratio of Joey's cards to Jimmy's?

4. A farm sells one ear of corn for every 3 tomatoes they sell. What is the ratio of tomatoes to corn?

5. At a restaurant, there are 4 chairs for every table. If there are 60 chairs, how many tables are there?

6. There are 20 students in Mr. Marshal's class. If the ratio of girls to boys is 3:1, how many girls and boys are in his class?

7. The ratio for the number of chocolate bars to lollipops is 4:2. If 24 lollipops were sold, how many chocolate bars were sold?

8. There are 30 flowers in Meg and Janet's garden. If the ratio of marigolds to roses is 2:1, how many of each flower do they have?

9. Ms. Jone's class surveyed 90 people and found that 3:6 people preferred water to soda. How many people preferred water and how many people preferred soda?

10. For every 5 computers sold, 3 people buy speakers. If 27 speakers were sold, how many computers were sold?

Ratios

mathbasics (1,2,3 ... to Pre-Algebra)   **Target time: 5 min   Student's time: ____ min**

Solve the following problems:

1. Sally has 50 grew roses. Stacy grew 35 roses. What is the ratio of Stacy's roses to Sally's?

2. Kendall had $66 at the beginning of the month and had $121 at the end of the month. What is the ratio of his money from the beginning of the month to the end?

3. Carol has 56 pairs of shoes and Jill has 64 pairs of shoes. What is the ratio of Jill's shoes to Carol's?

4. An aquarium has two starfish for every 10 goldfish. What is the ratio of starfish to goldfish?

5. In an apartment complex, there are 30 apartments to ever building. If there are 150 apartments, how many buildings are there?

6. There were 32 pizzas sold at The Pizza Place. If the ratio of pepperoni pizza to cheese is 5:3, how many pepperoni and cheese pizzas were ordered?

7. The ratio for the number of chickens to cows on a farm is 3:4. If there are 42 animals on the farm, how many of each animal are there?

8. There are 55 pairs of socks in Maggie's sock drawer. If the ratio of ankle socks to normal socks is 3:8, how many of each kind of sock is there?

9. Mrs. Jackson's class has 45 kids. If the ratio of boys to girls is 5:4, how many girls and boys are in the class?

10. For every 7 customers at a store, one person gets a free gift. If 54 gifts were given out, how many customers came to the store?

Ratios

Unit: 50
Worksheet: 3

mathbasics (1,2,3 ... to Pre-Algebra) **Target time: 5 min** **Student's time: ____ min**

Solve the following problems:

1. Fran has collected 20 key chains. Felicia has collected 35 key chains. What is the ratio of Fran's roses to Felicia's?

2. Carrie had a $50 gift certificate and spent $40 dollars of it. What is the ratio of the amount of money she has left to her original amount?

3. Sam grilled 150 hotdogs at the grilling tournament and Joe grilled 75. What is the ratio of Joe's hotdogs to Sam's?

4. A dentist office gives two packets of floss for every two toothbrushes they give out. What is the ratio of floss to toothbrushes?

5. A credit card company gives one reward point for every $10 spent on their card. If a person has 450 reward points, how much money have they spent?

6. There were 56 raffle tickets in a bucket. If the ratio of winners to non-winners was 2:5, how many winners and non-winners were there?

7. A store had a HUGE closeout sale and the ratio of new customers to regulars was 5:2. If there were 567 customers, how many of each customer was there?

8. There are 36 workers in a factory. If the ratio of the packers to machine operators in 2:4, how many of each kind of worker is there?

9. Mr. Wheeler's class has 27 kids. If the ratio of people wearing shorts to pants is 1:2, how many people are wearing shorts and how many are wearing pants?

10. Bobby gave out 2 apples for every 9 people that went to the health fair. If 108 people came to the health fair, how many apples were given out?

Proportions

Unit: 51
Worksheet: 1

mathbasics (1,2,3 ... to Pre-Algebra) **Target time: 3 min Student's time: ____ min**

Solve the following problems:

3:9 :: 5:?

2:4 :: 7:?

5:? :: 6:42

9:? :: 3:12

If 3 birds lay 15 eggs, how many birds would it take to lay 40 eggs?

If 10 people can eat 8 cakes, how many people would it take to eat 20 cakes?

If 5 boys can lift 100 pounds, how many boys would it take to lift 40 pounds?

If 4 bees pollinate 10 flowers, how many flowers can 8 bees pollinate?

If 8 girl scouts sell 32 cookies, how many cookies can 15 girl scouts sell?

If 3 people make 12 jackets, how many jackets can 20 people make?

Solve the following problems:

5:40 :: 9:?

2:24 :: 3:?

7:? :: 2:36

12:? :: 16:64

If 2 machines make 40 toys, how many machines would it take to make 100 toys?

If 3 people can make 18 candy bars, how many people would it take to make 60 candy bars?

If 4 horses can pull 3 carriages, how many horses would it take to pull 27 carriages?

If 7 people washed 21 towels, how many towels can 4 people wash?

If 24 coins are worth $25, how much is 48 coins worth?

If 12 kids made 4 pies, how many pies can 90 kids make?

Proportions

Unit: 51
Worksheet: 3

mathbasics (1,2,3 ... to Pre-Algebra) **Target time: 3 min** **Student's time: ____ min**

Solve the following problems:

3:9 :: 5:?

15:3 :: ?:8

4:? :: 6:12

9:? :: 3:12

If 3 birds lay 15 eggs, how many birds would it take to lay 40 eggs?

If 3 people can paint 2 houses in a day, how many people would it take to paint 10 houses?

If 6 plants produce 10 vegetables, how many plants would it take to grow 25 vegetables?

If 4 girls build 10 birdhouses, how many birdhouses can 14 girls make?

If 6 fish eat 12 worms, how many worms can 17 fish eat?

If 2 jars of jelly fill 14 packets, how many packets can 7 jars fill?

Signed Numbers, Addition

Unit: 52
Worksheet: 1

mathbasics (1,2,3 ... to Pre-Algebra) **Target time: 3 min** **Student's time: ____ min**

Solve the following problems:

25 + 42 =	-52 + 48 =	28 + 40 =
-14 + 91 =	86 - 39 =	-10 + 29 =
73 + 46 =	50 - 45 =	23 + 52 =
-94 + 12 =	25 - 19 =	-35 - 57 =
-44 + 59 =	-36 + 100 =	80 - 39 =
87 + 90 =	82 + 94 =	-68 + 84 =

Signed Numbers, Addition

Unit: 52
Worksheet: 2

mathbasics (1,2,3 ... to Pre-Algebra) **Target time: 3 min Student's time: ____ min**

Solve the following problems:

-39 - 84 =	-76 - 32 =	-26 - 71 =
-17 - 53 =	20 + 94 =	38 - 4 =
86 - 90 =	-8 - 36 =	-66 - 47 =
-58 - 72 =	48 + 43 =	39 - 2 =
14 + 74 =	6 - 37 =	-30 + 99 =
-96 + 4 =	94 + 49 =	-52 + 38 =

Signed Numbers, Addition

Unit: 52
Worksheet: 3

mathbasics (1,2,3 ... to Pre-Algebra) **Target time: 3 min** **Student's time: ____ min**

Solve the following problems:

-61 - 84 =	-27 - 74 =	36 - 15 =
-55 - 71 =	-80 - 99 =	-67 + 3 =
34 + 8 =	-79 - 42 =	58 - 79 =
77 + 60 =	-93 - 42 =	60 + 83 =
-28 + 14 =	-4 + 13 =	5 + 91 =
91 + 3 =	64 + 78 =	91 - 28 =

Signed Numbers, Subtraction

Unit: 53
Worksheet: 1

mathbasics (1,2,3 ... to Pre-Algebra) **Target time: 3 min Student's time: ____ min**

Solve the following problems:

44 - 41 =	-7 - (-8) =	-38 - (-60) =
-27 - 56 =	5 - (-45) =	-54 - (-62) =
9 - 38 =	63 - (-56) =	-28 - 59 =
-74 - 28 =	-58 - (-70) =	7 - (-54) =
-37 - (-7) =	-98 - 63 =	57 - 34 =
61 - (-3) =	74 - (-30) =	-59 - (-66) =

Signed Numbers, Subtraction

Unit: 53
Worksheet: 2

mathbasics (1,2,3 ... to Pre-Algebra) **Target time: 3 min** **Student's time: ____ min**

Solve the following problems:

-25 - (-23) =	33 - (-28) =	-21 - 66 =
-84 - (-74) =	-51 - (-92) =	-16 - (-15) =
33 - (-80) =	3 - 12 =	-43 - (-12) =
-71 - 95 =	-34 - 43 =	-21 - (-45) =
9 - 94 =	-63 - (-4) =	-14 - (-96) =
-73 - 11 =	37 - (-82) =	28 - (-72) =

Signed Numbers, Subtraction

Unit: 53
Worksheet: 3

mathbasics (1,2,3 ... to Pre-Algebra) **Target time: 3 min Student's time: ____ min**

Solve the following problems:

82 - 44 =	-67 - 25 =	8 - 72 =
75 - (-31) =	-83 - (-35) =	-17 - (-68) =
12 - (-23) =	-96 - (-64) =	-64 - 38 =
69 - 61 =	-31 - (-25) =	-1 - 92 =
62 - (-2) =	85 - 27 =	-71 - 21 =
-10 - 33 =	-8 - 46 =	-73 - 40 =

Signed Numbers, Multiplication

Unit: 54
Worksheet: 1

mathbasics (1,2,3 ... to Pre-Algebra) **Target time: 8 min** **Student's time: ____ min**

Solve the following problems:

$25 \times 73 =$	$-28 \times 29 =$	$-75 \times 26 =$
$-5 \times -33 =$	$-55 \times -89 =$	$-48 \times 94 =$
$19 \times -16 =$	$99 \times 11 =$	$46 \times 14 =$
$-65 \times -24 =$	$-27 \times 14 =$	$32 \times 44 =$
$-56 \times 23 =$	$89 \times 79 =$	$84 \times -72 =$
$51 \times 1 =$	$-61 \times -51 =$	$80 \times 94 =$

Signed Numbers, Multiplication

Unit: 54
Worksheet: 2

mathbasics (1,2,3 ... to Pre-Algebra) **Target time: 8 min Student's time: ____ min**

Solve the following problems:

$-27 \times -41 =$	$-37 \times -45 =$	$-97 \times 69 =$
$-18 \times -97 =$	$94 \times 52 =$	$29 \times 25 =$
$-9 \times 23 =$	$-29 \times 82 =$	$23 \times -32 =$
$65 \times 65 =$	$-68 \times -81 =$	$25 \times -7 =$
$-31 \times -15 =$	$-63 \times -29 =$	$-65 \times 30 =$
$26 \times 92 =$	$70 \times -3 =$	$68 \times -16 =$

Signed Numbers, Multiplication

Unit: 54
Worksheet: 3

mathbasics (1,2,3 ... to Pre-Algebra) **Target time: 8 min** **Student's time: ____ min**

Solve the following problems:

$-10 \times 94 =$	$59 \times 23 =$	$86 \times 12 =$
$-71 \times -20 =$	$-41 \times 14 =$	$-90 \times -76 =$
$-89 \times 55 =$	$-51 \times -22 =$	$88 \times -39 =$
$-26 \times -80 =$	$65 \times 64 =$	$9 \times 38 =$
$96 \times 71 =$	$-59 \times 30 =$	$-5 \times -4 =$
$-61 \times -40 =$	$85 \times -58 =$	$-88 \times 60 =$

Signed Numbers, Division

Unit: 55
Worksheet: 1

mathbasics (1,2,3 ... to Pre-Algebra) **Target time: 6 min** **Student's time: ____ min**

Solve the following problems:

144 ÷ 48 =	-3 ÷ -1 =	-750 ÷ 75 =
-322 ÷ -46 =	-910 ÷ 91 =	0 ÷ 78 =
0 ÷ 57 =	-282 ÷ 47 =	950 ÷ 95 =
-90 ÷ -9 =	160 ÷ -20 =	-198 ÷ -99 =
474 ÷ 79 =	864 ÷ -96 =	-540 ÷ -60 =
-56 ÷ -28 =	124 ÷ -62 =	-445 ÷ 89 =

Signed Numbers, Division

Unit: 55
Worksheet: 2

mathbasics (1,2,3 ... to Pre-Algebra)   **Target time: 6 min   Student's time: ____ min**

Solve the following problems:

312 ÷ -39 =	201 ÷ 67 =	200 ÷ -50 =
126 ÷ 14 =	-188 ÷ 94 =	-240 ÷ 48 =
198 ÷ 66 =	-580 ÷ 58 =	-166 ÷ -83 =
-432 ÷ -72 =	160 ÷ 16 =	10 ÷ -2 =
-400 ÷ 40 =	-66 ÷ -22 =	44 ÷ 11 =
840 ÷ -84 =	20 ÷ -4 =	-128 ÷ 16 =

Signed Numbers, Division

Unit: 55
Worksheet: 3

mathbasics (1,2,3 ... to Pre-Algebra) **Target time: 6 min** **Student's time: ____ min**

Solve the following problems:

-74 ÷ 37 =	-98 ÷ 98 =	250 ÷ -50 =
0 ÷ -84 =	-200 ÷ -40 =	355 ÷ -71 =
42 ÷ -6 =	-837 ÷ -93 =	-23 ÷ 23 =
-36 ÷ 4 =	240 ÷ 48 =	276 ÷ 69 =
265 ÷ -53 =	140 ÷ 14 =	-160 ÷ -40 =
231 ÷ 33 =	-231 ÷ 33 =	-590 ÷ 59 =

Absolute Value

Unit: 56
Worksheet: 1

mathbasics (1,2,3 ... to Pre-Algebra) **Target time: 4 min** **Student's time: ____ min**

Solve the following problems:

$\|-83\| =$	$\|-78\| =$	$\|38\| =$	$\|-13\| =$
$\|-7\| =$	$\|1\| =$	$\|-67\| =$	$\|-46\| =$
$\|-25\| =$	$\|48\| =$	$\|60\| =$	$\|26\| =$

$\|-213 \div -71\| =$	$\|17 + 27\| =$	$\|-2 + 48\| =$
$\|36 - 36\| =$	$\|3 - 64\| =$	$\|-59 \times 33\| =$
$\|-57 + 2\| =$	$\|328 \div 82\| =$	$\|-324 \div 54\| =$
$\|-83 \times 55\| =$	$\|89 + (-23)\| =$	$\|-588 \div -84\| =$

Absolute Value

Unit: 56
Worksheet: 2

mathbasics (1,2,3 ... to Pre-Algebra)

Target time: 4 min **Student's time: ____ min**

Solve the following problems:

$| -42 | =$ $| 49 | =$ $| -91 | =$ $| 20 | =$

$| -51 | =$ $| -34 | =$ $| -64 | =$ $| 53 | =$

$| -44 | =$ $| 75 | =$ $| 74 | =$ $| 99 | =$

$| -288 \div 96 | =$ $| 20 - (-95) | =$ $| 82 \times 41 | =$

$| 236 \div -59 | =$ $| -20 \times -87 | =$ $| 60 + (-30) | =$

$| 6 \times -28 | =$ $| 61 + 28 | =$ $| -54 + (-62) | =$

$| -31 \times -58 | =$ $| 55 + (-99) | =$ $| 73 \times -50 | =$

Absolute Value

Unit: 56
Worksheet: 3

mathbasics (1,2,3 ... to Pre-Algebra) **Target time: 4 min** **Student's time: ____ min**

Solve the following problems:

$\| 51 \| =$	$\| -66 \| =$	$\| -34 \| =$	$\| -90 \| =$
$\| 94 \| =$	$\| 80 \| =$	$\| -44 \| =$	$\| -87 \| =$
$\| 15 \| =$	$\| -21 \| =$	$\| -85 \| =$	$\| -98 \| =$

$\| -124 \div 62 \| =$	$\| -60 \times -56 \| =$	$\| 0 \div 30 \| =$
$\| 74 - 26 \| =$	$\| -38 \div -38 \| =$	$\| -480 \div -48 \| =$
$\| -22 \times -24 \| =$	$\| 70 + (-64) \| =$	$\| -36 \times 20 \| =$
$\| -68 - 9 \| =$	$\| 396 \div 66 \| =$	$\| 14 \times 89 \| =$

Answer Key to the Assessment Test

1. 8
2. 4
3. 3, 000, 000
4. 9
5. 86
6. 237
7. 12835
8. 1
9. 2015
10. 3318
11. 4429
12. 84
13. 56
14. 316
15. 6204
16. 39186
17. 2730043
18. 2r2
19. 100r3
20. 223r14
21. 48
22. 625
23. $8\frac{58}{67}$
24. $\frac{5892}{366}$
25. $2^3 \times 3^2 \times 5 \times 7^2$
26. 1715
27. $\frac{3}{5}$
28. $\frac{634}{33}$
29. $\frac{13}{8}$
30. $\frac{25049}{3430}$

Answer Key to the Assessment Test (continued)

31. $\frac{16}{3}$
32. $\frac{28}{27}$
33. $\frac{3}{1250}$
34. $\frac{24}{65}$
35. 56+(82+42)
36. (73×24)+(73×52)
37. 252
38. 0.00018
39. 802.187918
40. 7326.08084
41. 4063.9667
42. 0.875
43. $160\frac{2}{5}$
44. 0.625
45. $\frac{325}{2}$
46. 56178.441%
47. 0.0532
48. 140%
49. 80
50. 45
51. 6
52. -4
53. 12
54. 3111
55. -2
56. 4

Counting Numbers

Unit: 01

Answers Worksheet: I

Targettime: **1** min Student's time: ____min

Cotmt the items:

7

8

1

3

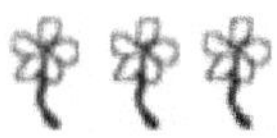

9

6

2

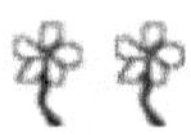

4

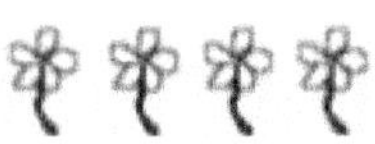

5

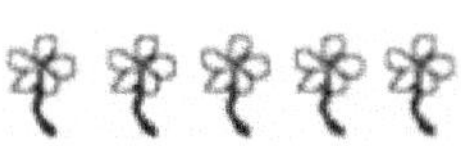

Draw the number of items indicated:

1

3

5

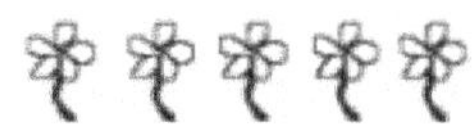

2

6

4

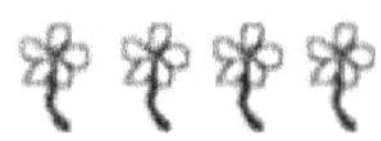

8

9

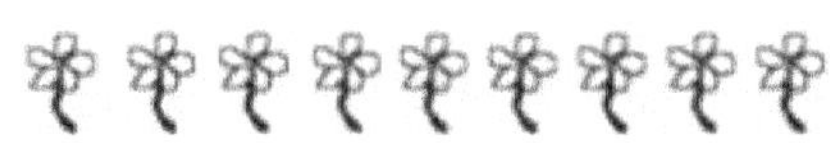

7

Counting Numbers

mathbasics {1,2,3 ... to Pre-Algebra}   Targettime: **1** min   Student's time: ____min

Cotmt the items:

8

2

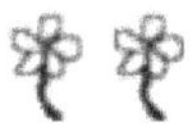

4

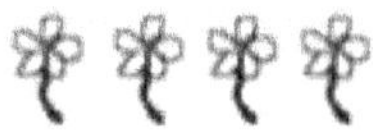

3

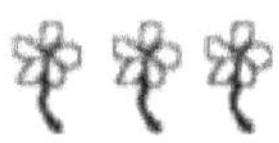

6

1

5

9

7

Draw the number of items indicated:

8

6

2

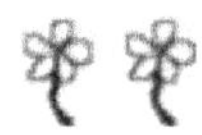

5

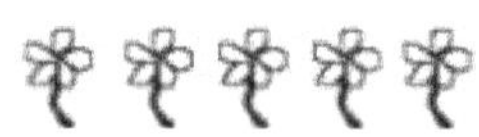

9

1

3

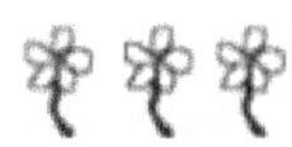

4

7

mathbasics {1,2,3 ... to Pre-Algebra} Targettime: 1 min Student's time: ____min

Cotmt the items:

5

8

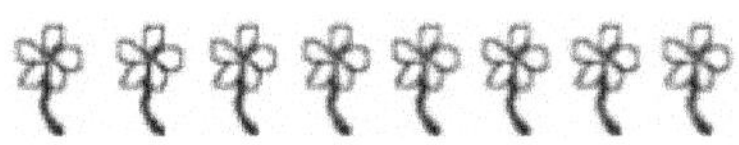

4

9

3

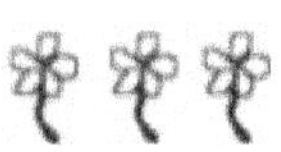

1

7

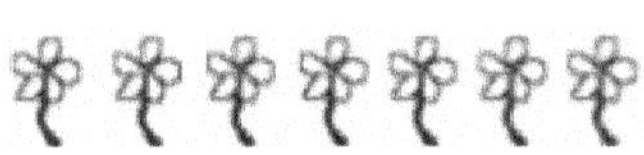

6

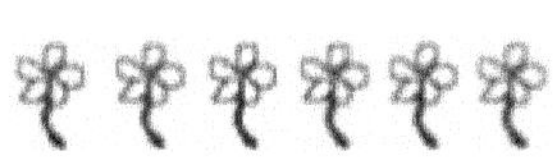

2

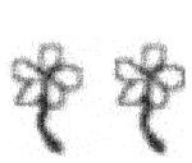

Draw the number of items indicated:

3

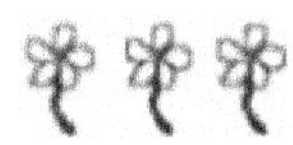

1

6

8

4

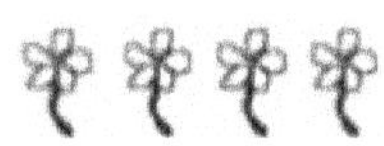

7

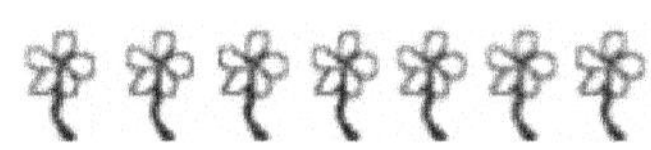

5

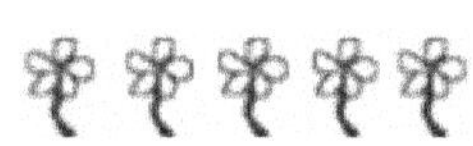

9

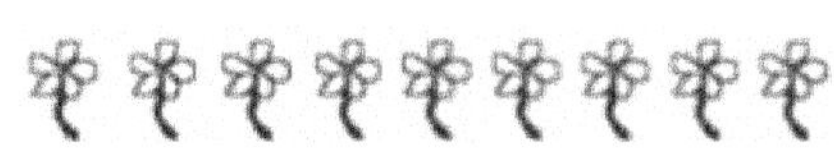

2

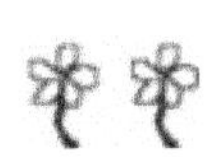

Target time: 5 min **Student's time: ____ min**

1. In 8,579,643, what is the place value of:

   7 - ten-thousands

   4 - tens

   8 - millions

   3 - ones

   9 - thousands

   5 - hundred-thousands

   6 - hundreds

2. In 2,810,596, what is the place value of:

   8 - hundred-thousands

   1 - ten-thousands

   9 - tens

   5 - hundreds

   6 - ones

   2 - millions

   0 - thousands

3. In 7,125,389, what is the place value of:

   1 - hundred-thousands

   8 - tens

   3 - hundreds

   7 - millions

   2 - ten-thousands

   5 - thousands

   9 - ones

4. In 6,349,582, what digit is in the:

   ones place - 2

   tens place - 8

   millions place - 6

   hundreds place - 5

   thousands place - 9

   ten-thousands place - 4

   hundred-thousands place - 3

5. In 9,265,341, what digit is in the:

   millions place - 9

   tens place - 4

   hundred-thousands place - 2

   hundreds place - 3

   ones place - 1

   ten-thousands place - 6

   thousands place - 5

6. In 8,192,057, what digit is in the:

   tens place - 5

   thousands place - 2

   millions place - 8

   ones place - 7

   hundreds place - 0

   hundred-thousands place - 1

   ten-thousands place - 9

Target time: 5 min **Student's time: ____ min**

1. In 3,865,290, what is the place value of:
   5 - thousands
   3 - millions
   6 - ten-thousands
   0 - ones
   8 - hundred-thousands
   9 - tens
   2 - hundreds

2. In 7,162,489, what is the place value of:
   2 - thousands
   8 - tens
   1 - hundred-thousands
   6 - ten-thousands
   7 - millions
   9 - ones
   4 - hundreds

3. In 9,702,483, what is the place value of:
   0 - ten-thousands
   7 - hundred-thousands
   4 - hundreds
   9 - millions
   2 - thousands
   8 - tens
   3 - ones

4. In 5,726,983, what digit is in the:
   tens place - 8
   ten-thousands place - 2
   ones place - 3
   millions place - 5
   hundreds place - 9
   hundred-thousands place - 7
   thousands place - 6

5. In 2,846,971, what digit is in the:
   ten-thousands place - 4
   ones place - 1
   millions place - 2
   thousands place - 6
   hundreds place - 9
   hundred-thousands place - 8
   tens place - 7

6. In 6,382,147, what digit is in the:
   thousands place - 2
   hundred-thousands place - 3
   tens place - 4
   millions place - 6
   hundreds place - 1
   ones place - 7
   ten-thousands place - 8

Target time: 5 min **Student's time: ____ min**

1. In 8,214,063, what is the place value of:

   2 - hundred-thousands

   1 - ten-thousands

   8 - millions

   4 - thousands

   3 - ones

   0 - hundreds

   6 - tens

2. In 8,309,421, what is the place value of:

   2 - tens

   3 - hundred-thousands

   4 - hundreds

   9 - thousands

   0 - ten-thousands

   8 - millions

   1 - ones

3. In 1,832,904, what is the place value of:

   4 - ones

   9 - hundreds

   0 - tens

   1 - millions

   3 - ten-thousands

   8 - hundred-thousands

   2 - thousands

4. In 9,645,378, what digit is in the:

   ten-thousands place - 4

   hundred-thousands place - 6

   tens place - 7

   thousands place - 5

   hundreds place - 3

   ones place - 8

   millions place - 9

5. In 5,023,748, what digit is in the:

   hundreds place - 7

   ten-thousands place - 2

   tens place - 4

   ones place - 8

   hundred-thousands place - 0

   thousands place - 3

   millions place - 5

6. In 4,581,739, what digit is in the:

   millions place - 4

   ones place - 9

   tens place - 3

   thousands place - 1

   ten-thousands place - 8

   hundreds place - 7

   hundred-thousands place - 5

Numbers 1-100 and Rounding

Unit: 03
Answers Worksheet: 1

mathbasics (1,2,3 ... to Pre-Algebra) **Target time: 5 min Student's time: ____ min**

Write each number in word form:

28 - twenty-eight	40 - fourty	14 - fourteen
15 - fifteen	2 - two	81 - eighty-one
20 - twenty	50 - fifty	22 - twenty-two
71 - seventy-one	25 - twenty-five	39 - thirty-nine
9 - nine	35 - thirty-five	75 - seventy-five
48 - fourty-eight	45 - fourty-five	16 - sixteen
37 - thirty-seven	58 - fifty-eight	7 - seven
5 - five	6 - six	42 - fourty-two
60 - sixty	84 - eighty-four	30 - thirty
93 - ninety-three	67 - sixty-seven	31 - thirty-one

Round each number to the designated place value:

9,852,702 (to ten-thousands) - 9,850,000	3,773,527 (to hundred-thousands) - 3,800,000
1,787,191 (to hundreds) - 1,787,200	7,706,633 (to thousands) - 7,707,000
8,782,124 (to millions) - 9,000,000	5,225,858 (to thousands) - 5,226,000
7,954,439 (to millions) - 8,000,000	7,747,684 (to tens) - 7,747,680

Write each number in word form:

70 - seventy	79 - seventy-nine	38 - thirty-eight
75 - seventy-five	51 - fifty-one	59 - fifty-nine
65 - sixty-five	8 - eight	78 - seventy-eight
15 - fifteen	99 - ninety-nine	9 - nine
23 - twenty-three	86 - eighty-six	67 - sixty-seven
54 - fifty-four	4 - four	2 - two
49 - fourty-nine	96 - ninety-six	61 - sixty-one
98 - ninety-eight	10 - ten	88 - eighty-eight
3 - three	17 - seventeen	47 - fourty-seven
89 - eighty-nine	27 - twenty-seven	55 - fifty-five

Round each number to the designated place value:

2,722,431 (to hundred-thousands) - 2,700,000	3,565,787 (to millions) - 4,000,000
3,609,912 (to hundreds) - 3,609,900	116,087 (to tens) - 116,090
4,245,029 (to ten-thousands) - 4,250,000	3,602,339 (to millions) - 4,000,000
8,382,844 (to thousands) - 8,383,000	3,816,726 (to thousands) - 3,817,000

Write each number in word form:

29 - twenty-nine	96 - ninety-six	33 - thirty-three
46 - fourty-six	34 - thirty-four	41 - fourty-one
20 - twenty	91 - ninety-one	95 - ninety-five
50 - fifty	77 - seventy-seven	13 - thirteen
40 - fourty	22 - twenty-two	12 - twelve
1 - one	48 - fourty-eight	25 - twenty-five
5 - five	18 - eighteen	17 - seventeen
52 - fifty-two	24 - twenty-four	57 - fifty-seven
45 - fourty-five	37 - thirty-seven	16 - sixteen
36 - thirty-six	53 - fifty-three	31 - thirty-one

Round each number to the designated place value:

6,226,440 (to hundred-thousands) - 6,200,000	7,906,701 (to tens) - 7,906,700
746,979 (to tens) - 746,980	3,999,500 (to hundreds) - 3,999,500
1,823,539 (to millions) - 2,000,000	7,870,323 (to ten-thousands) - 7,870,000
8,608,462 (to hundred-thousands) - 8,600,000	4,097,709 (to thousands) - 4,098,000

Solve the following problems:

$$\begin{array}{r} 4 \\ +2 \\ \hline 6 \end{array} \qquad \begin{array}{r} 4 \\ +4 \\ \hline 8 \end{array} \qquad \begin{array}{r} 4 \\ +5 \\ \hline 9 \end{array}$$

$$\begin{array}{r} 3 \\ +1 \\ \hline 4 \end{array} \qquad \begin{array}{r} 7 \\ +1 \\ \hline 8 \end{array} \qquad \begin{array}{r} 2 \\ +7 \\ \hline 9 \end{array}$$

$$\begin{array}{r} 1 \\ +6 \\ \hline 7 \end{array} \qquad \begin{array}{r} 3 \\ +1 \\ \hline 4 \end{array} \qquad \begin{array}{r} 4 \\ +1 \\ \hline 5 \end{array}$$

$$\begin{array}{r} 6 \\ 1 \\ +2 \\ \hline 9 \end{array} \qquad \begin{array}{r} 6 \\ 2 \\ +1 \\ \hline 9 \end{array} \qquad \begin{array}{r} 1 \\ 7 \\ +1 \\ \hline 9 \end{array}$$

$$\begin{array}{r} 6 \\ 1 \\ +1 \\ \hline 8 \end{array} \qquad \begin{array}{r} 2 \\ 4 \\ +3 \\ \hline 9 \end{array} \qquad \begin{array}{r} 6 \\ 1 \\ +1 \\ \hline 8 \end{array}$$

Solve the following problems:

$$\begin{array}{r} 5 \\ +3 \\ \hline 8 \end{array} \qquad \begin{array}{r} 2 \\ +3 \\ \hline 5 \end{array} \qquad \begin{array}{r} 1 \\ +4 \\ \hline 5 \end{array}$$

$$\begin{array}{r} 4 \\ +1 \\ \hline 5 \end{array} \qquad \begin{array}{r} 2 \\ +5 \\ \hline 7 \end{array} \qquad \begin{array}{r} 2 \\ +4 \\ \hline 6 \end{array}$$

$$\begin{array}{r} 7 \\ +1 \\ \hline 8 \end{array} \qquad \begin{array}{r} 1 \\ +3 \\ \hline 4 \end{array} \qquad \begin{array}{r} 7 \\ +2 \\ \hline 9 \end{array}$$

$$\begin{array}{r} 7 \\ 1 \\ +1 \\ \hline 9 \end{array} \qquad \begin{array}{r} 1 \\ 6 \\ +2 \\ \hline 9 \end{array} \qquad \begin{array}{r} 4 \\ 1 \\ +2 \\ \hline 7 \end{array}$$

$$\begin{array}{r} 3 \\ 5 \\ +1 \\ \hline 9 \end{array} \qquad \begin{array}{r} 4 \\ 2 \\ +1 \\ \hline 7 \end{array} \qquad \begin{array}{r} 5 \\ 3 \\ +1 \\ \hline 9 \end{array}$$

Solve the following problems:

$$\begin{array}{r} 1 \\ +3 \\ \hline 4 \end{array} \qquad \begin{array}{r} 3 \\ +2 \\ \hline 5 \end{array} \qquad \begin{array}{r} 8 \\ +1 \\ \hline 9 \end{array}$$

$$\begin{array}{r} 7 \\ +2 \\ \hline 9 \end{array} \qquad \begin{array}{r} 5 \\ +4 \\ \hline 9 \end{array} \qquad \begin{array}{r} 8 \\ +1 \\ \hline 9 \end{array}$$

$$\begin{array}{r} 4 \\ +5 \\ \hline 9 \end{array} \qquad \begin{array}{r} 4 \\ +5 \\ \hline 9 \end{array} \qquad \begin{array}{r} 1 \\ +2 \\ \hline 3 \end{array}$$

$$\begin{array}{r} 4 \\ 3 \\ +1 \\ \hline 8 \end{array} \qquad \begin{array}{r} 6 \\ 2 \\ +1 \\ \hline 9 \end{array} \qquad \begin{array}{r} 3 \\ 2 \\ +3 \\ \hline 8 \end{array}$$

$$\begin{array}{r} 4 \\ 1 \\ +3 \\ \hline 8 \end{array} \qquad \begin{array}{r} 5 \\ 1 \\ +3 \\ \hline 9 \end{array} \qquad \begin{array}{r} 6 \\ 1 \\ +2 \\ \hline 9 \end{array}$$

Addition, Level 2

Unit: 05
Answers Worksheet: 1

mathbasics (1,2,3 ... to Pre-Algebra) **Target time: 1 min Student's time: ____ min**

Solve the following problems:

$$\begin{array}{r} 63 \\ +11 \\ \hline 74 \end{array} \qquad \begin{array}{r} 11 \\ +71 \\ \hline 82 \end{array} \qquad \begin{array}{r} 20 \\ +40 \\ \hline 60 \end{array}$$

$$\begin{array}{r} 58 \\ +20 \\ \hline 78 \end{array} \qquad \begin{array}{r} 77 \\ +10 \\ \hline 87 \end{array} \qquad \begin{array}{r} 25 \\ +51 \\ \hline 76 \end{array}$$

$$\begin{array}{r} 17 \\ +51 \\ \hline 68 \end{array} \qquad \begin{array}{r} 77 \\ +12 \\ \hline 89 \end{array} \qquad \begin{array}{r} 11 \\ +12 \\ \hline 23 \end{array}$$

$$\begin{array}{r} 13 \\ 21 \\ +52 \\ \hline 86 \end{array} \qquad \begin{array}{r} 38 \\ 20 \\ +31 \\ \hline 89 \end{array} \qquad \begin{array}{r} 41 \\ 26 \\ +12 \\ \hline 79 \end{array}$$

$$\begin{array}{r} 73 \\ 15 \\ +10 \\ \hline 98 \end{array} \qquad \begin{array}{r} 61 \\ 18 \\ +10 \\ \hline 89 \end{array} \qquad \begin{array}{r} 24 \\ 23 \\ +30 \\ \hline 77 \end{array}$$

mathbasics (1,2,3 ... to Pre-Algebra) **Target time: 1 min** **Student's time: ____ min**

Solve the following problems:

$$\begin{array}{r} 16 \\ +63 \\ \hline 79 \end{array} \qquad \begin{array}{r} 77 \\ +11 \\ \hline 88 \end{array} \qquad \begin{array}{r} 31 \\ +54 \\ \hline 85 \end{array}$$

$$\begin{array}{r} 55 \\ +14 \\ \hline 69 \end{array} \qquad \begin{array}{r} 72 \\ +15 \\ \hline 87 \end{array} \qquad \begin{array}{r} 26 \\ +31 \\ \hline 57 \end{array}$$

$$\begin{array}{r} 55 \\ +30 \\ \hline 85 \end{array} \qquad \begin{array}{r} 44 \\ +13 \\ \hline 57 \end{array} \qquad \begin{array}{r} 64 \\ +14 \\ \hline 78 \end{array}$$

$$\begin{array}{r} 22 \\ 24 \\ +11 \\ \hline 57 \end{array} \qquad \begin{array}{r} 15 \\ 63 \\ +20 \\ \hline 98 \end{array} \qquad \begin{array}{r} 51 \\ 17 \\ +21 \\ \hline 89 \end{array}$$

$$\begin{array}{r} 27 \\ 12 \\ +20 \\ \hline 59 \end{array} \qquad \begin{array}{r} 67 \\ 21 \\ +10 \\ \hline 98 \end{array} \qquad \begin{array}{r} 33 \\ 54 \\ +11 \\ \hline 98 \end{array}$$

mathbasics (1,2,3 ... to Pre-Algebra) **Target time: 1 min** **Student's time: ____ min**

Solve the following problems:

$$\begin{array}{r} 12 \\ +55 \\ \hline 67 \end{array} \qquad \begin{array}{r} 21 \\ +67 \\ \hline 88 \end{array} \qquad \begin{array}{r} 76 \\ +11 \\ \hline 87 \end{array}$$

$$\begin{array}{r} 25 \\ +54 \\ \hline 79 \end{array} \qquad \begin{array}{r} 42 \\ +42 \\ \hline 84 \end{array} \qquad \begin{array}{r} 58 \\ +30 \\ \hline 88 \end{array}$$

$$\begin{array}{r} 14 \\ +22 \\ \hline 36 \end{array} \qquad \begin{array}{r} 78 \\ +10 \\ \hline 88 \end{array} \qquad \begin{array}{r} 47 \\ +12 \\ \hline 59 \end{array}$$

$$\begin{array}{r} 16 \\ 70 \\ +10 \\ \hline 96 \end{array} \qquad \begin{array}{r} 40 \\ 21 \\ +24 \\ \hline 85 \end{array} \qquad \begin{array}{r} 17 \\ 52 \\ +20 \\ \hline 89 \end{array}$$

$$\begin{array}{r} 29 \\ 40 \\ +10 \\ \hline 79 \end{array} \qquad \begin{array}{r} 48 \\ 40 \\ +11 \\ \hline 99 \end{array} \qquad \begin{array}{r} 74 \\ 13 \\ +12 \\ \hline 99 \end{array}$$

mathbasics (1,2,3 ... to Pre-Algebra) Target time: 2 min Student's time: ____ min

Solve the following problems:

$$\begin{array}{r} 39 \\ +68 \\ \hline 107 \end{array} \qquad \begin{array}{r} 33 \\ +99 \\ \hline 132 \end{array} \qquad \begin{array}{r} 94 \\ +66 \\ \hline 160 \end{array}$$

$$\begin{array}{r} 58 \\ +67 \\ \hline 125 \end{array} \qquad \begin{array}{r} 57 \\ +54 \\ \hline 111 \end{array} \qquad \begin{array}{r} 15 \\ +49 \\ \hline 64 \end{array}$$

$$\begin{array}{r} 61 \\ +29 \\ \hline 90 \end{array} \qquad \begin{array}{r} 73 \\ +57 \\ \hline 130 \end{array} \qquad \begin{array}{r} 28 \\ +25 \\ \hline 53 \end{array}$$

$$\begin{array}{r} 22 \\ 18 \\ +85 \\ \hline 125 \end{array} \qquad \begin{array}{r} 23 \\ 29 \\ +17 \\ \hline 69 \end{array} \qquad \begin{array}{r} 69 \\ 76 \\ +24 \\ \hline 169 \end{array}$$

$$\begin{array}{r} 56 \\ 87 \\ +94 \\ \hline 237 \end{array} \qquad \begin{array}{r} 24 \\ 27 \\ +52 \\ \hline 103 \end{array} \qquad \begin{array}{r} 75 \\ 69 \\ +72 \\ \hline 216 \end{array}$$

Solve the following problems:

$$\begin{array}{r} 45 \\ +98 \\ \hline 143 \end{array} \quad \begin{array}{r} 84 \\ +57 \\ \hline 141 \end{array} \quad \begin{array}{r} 41 \\ +39 \\ \hline 80 \end{array}$$

$$\begin{array}{r} 84 \\ +57 \\ \hline 141 \end{array} \quad \begin{array}{r} 28 \\ +34 \\ \hline 62 \end{array} \quad \begin{array}{r} 66 \\ +24 \\ \hline 90 \end{array}$$

$$\begin{array}{r} 67 \\ +33 \\ \hline 100 \end{array} \quad \begin{array}{r} 33 \\ +59 \\ \hline 92 \end{array} \quad \begin{array}{r} 44 \\ +87 \\ \hline 131 \end{array}$$

$$\begin{array}{r} 47 \\ 78 \\ +27 \\ \hline 152 \end{array} \quad \begin{array}{r} 33 \\ 99 \\ +44 \\ \hline 176 \end{array} \quad \begin{array}{r} 17 \\ 27 \\ +63 \\ \hline 107 \end{array}$$

$$\begin{array}{r} 76 \\ 16 \\ +28 \\ \hline 120 \end{array} \quad \begin{array}{r} 96 \\ 45 \\ +49 \\ \hline 190 \end{array} \quad \begin{array}{r} 67 \\ 75 \\ +58 \\ \hline 200 \end{array}$$

mathbasics (1,2,3 ... to Pre-Algebra) Target time: 2 min Student's time: ____ min

Solve the following problems:

$$\begin{array}{r} 28 \\ +85 \\ \hline 113 \end{array} \qquad \begin{array}{r} 68 \\ +27 \\ \hline 95 \end{array} \qquad \begin{array}{r} 53 \\ +57 \\ \hline 110 \end{array}$$

$$\begin{array}{r} 93 \\ +98 \\ \hline 191 \end{array} \qquad \begin{array}{r} 45 \\ +46 \\ \hline 91 \end{array} \qquad \begin{array}{r} 96 \\ +05 \\ \hline 101 \end{array}$$

$$\begin{array}{r} 91 \\ +29 \\ \hline 120 \end{array} \qquad \begin{array}{r} 65 \\ +58 \\ \hline 123 \end{array} \qquad \begin{array}{r} 42 \\ +38 \\ \hline 80 \end{array}$$

$$\begin{array}{r} 67 \\ 04 \\ +60 \\ \hline 131 \end{array} \qquad \begin{array}{r} 31 \\ 59 \\ +14 \\ \hline 104 \end{array} \qquad \begin{array}{r} 17 \\ 59 \\ +64 \\ \hline 140 \end{array}$$

$$\begin{array}{r} 37 \\ 54 \\ +23 \\ \hline 114 \end{array} \qquad \begin{array}{r} 82 \\ 78 \\ +23 \\ \hline 183 \end{array} \qquad \begin{array}{r} 34 \\ 89 \\ +10 \\ \hline 133 \end{array}$$

Addition, Level 4

Unit: 07
Answers Worksheet: 1

mathbasics (1,2,3 ... to Pre-Algebra) **Target time: 4 min** **Student's time: ____ min**

Solve the following problems:

$$\begin{array}{r} 9499 \\ +2546 \\ \hline 12045 \end{array} \qquad \begin{array}{r} 1911 \\ +9489 \\ \hline 11400 \end{array} \qquad \begin{array}{r} 7333 \\ +5609 \\ \hline 12942 \end{array}$$

$$\begin{array}{r} 5448 \\ +6736 \\ \hline 12184 \end{array} \qquad \begin{array}{r} 2345 \\ +9805 \\ \hline 12150 \end{array} \qquad \begin{array}{r} 7207 \\ +6985 \\ \hline 14192 \end{array}$$

$$\begin{array}{r} 3168 \\ +7664 \\ \hline 10832 \end{array} \qquad \begin{array}{r} 1117 \\ +8523 \\ \hline 9640 \end{array} \qquad \begin{array}{r} 9658 \\ +4583 \\ \hline 14241 \end{array}$$

$$\begin{array}{r} 8995 \\ 9886 \\ +9407 \\ \hline 28288 \end{array} \qquad \begin{array}{r} 1702 \\ 8289 \\ +8087 \\ \hline 18078 \end{array} \qquad \begin{array}{r} 7362 \\ 9549 \\ +5315 \\ \hline 22226 \end{array}$$

$$\begin{array}{r} 2972 \\ 4799 \\ +5064 \\ \hline 12835 \end{array} \qquad \begin{array}{r} 3501 \\ 5989 \\ +2442 \\ \hline 11932 \end{array} \qquad \begin{array}{r} 9894 \\ 1197 \\ +645 \\ \hline 11736 \end{array}$$

Solve the following problems:

4309 +3726 8035	3378 +2506 5884	9579 +3635 13214
7288 +8334 15622	7321 +1479 8800	8412 +268 8680
7492 +369 7861	6465 +8077 14542	4965 +4486 9451
9049 2101 +5916 17066	3849 2032 +165 6046	2306 7897 +8452 18655
6513 5389 +1275 13177	4155 3297 +5913 13365	8187 6484 +46 14717

Solve the following problems:

$$\begin{array}{r} 4345 \\ +6105 \\ \hline 10450 \end{array}$$

$$\begin{array}{r} 3546 \\ +8416 \\ \hline 11962 \end{array}$$

$$\begin{array}{r} 4263 \\ +4597 \\ \hline 8860 \end{array}$$

$$\begin{array}{r} 9739 \\ +7852 \\ \hline 17591 \end{array}$$

$$\begin{array}{r} 2978 \\ +1785 \\ \hline 4763 \end{array}$$

$$\begin{array}{r} 3302 \\ +7248 \\ \hline 10550 \end{array}$$

$$\begin{array}{r} 5149 \\ +6396 \\ \hline 11545 \end{array}$$

$$\begin{array}{r} 1258 \\ +4 \\ \hline 1262 \end{array}$$

$$\begin{array}{r} 9711 \\ +4949 \\ \hline 14660 \end{array}$$

$$\begin{array}{r} 3581 \\ 7329 \\ +4711 \\ \hline 15621 \end{array}$$

$$\begin{array}{r} 1162 \\ 4208 \\ +2072 \\ \hline 7442 \end{array}$$

$$\begin{array}{r} 7822 \\ 8798 \\ +2721 \\ \hline 19341 \end{array}$$

$$\begin{array}{r} 2436 \\ 1475 \\ +5439 \\ \hline 9350 \end{array}$$

$$\begin{array}{r} 6348 \\ 4203 \\ +1239 \\ \hline 11790 \end{array}$$

$$\begin{array}{r} 9303 \\ 1237 \\ +7689 \\ \hline 18229 \end{array}$$

Subtraction, Level 1

Unit: 08
Answers Worksheet: 1

mathbasics (1,2,3 ... to Pre-Algebra) **Target time: 1 min** **Student's time: ____ min**

Solve the following problems:

$$\begin{array}{r} 9 \\ -5 \\ \hline 4 \end{array} \qquad \begin{array}{r} 6 \\ -2 \\ \hline 4 \end{array} \qquad \begin{array}{r} 1 \\ -1 \\ \hline 0 \end{array}$$

$$\begin{array}{r} 4 \\ -3 \\ \hline 1 \end{array} \qquad \begin{array}{r} 4 \\ -4 \\ \hline 0 \end{array} \qquad \begin{array}{r} 3 \\ -2 \\ \hline 1 \end{array}$$

$$\begin{array}{r} 9 \\ -7 \\ \hline 2 \end{array} \qquad \begin{array}{r} 5 \\ -5 \\ \hline 0 \end{array} \qquad \begin{array}{r} 7 \\ -3 \\ \hline 4 \end{array}$$

$$\begin{array}{r} 9 \\ -5 \\ \hline 4 \end{array} \qquad \begin{array}{r} 3 \\ -1 \\ \hline 2 \end{array} \qquad \begin{array}{r} 7 \\ -6 \\ \hline 1 \end{array}$$

$$\begin{array}{r} 2 \\ -1 \\ \hline 1 \end{array} \qquad \begin{array}{r} 3 \\ -3 \\ \hline 0 \end{array} \qquad \begin{array}{r} 7 \\ -4 \\ \hline 3 \end{array}$$

Solve the following problems:

$$\begin{array}{r} 2 \\ -2 \\ \hline 0 \end{array} \qquad \begin{array}{r} 9 \\ -8 \\ \hline 1 \end{array} \qquad \begin{array}{r} 8 \\ -1 \\ \hline 7 \end{array}$$

$$\begin{array}{r} 3 \\ -1 \\ \hline 2 \end{array} \qquad \begin{array}{r} 6 \\ -5 \\ \hline 1 \end{array} \qquad \begin{array}{r} 2 \\ -2 \\ \hline 0 \end{array}$$

$$\begin{array}{r} 2 \\ -1 \\ \hline 1 \end{array} \qquad \begin{array}{r} 6 \\ -4 \\ \hline 2 \end{array} \qquad \begin{array}{r} 9 \\ -4 \\ \hline 5 \end{array}$$

$$\begin{array}{r} 9 \\ -9 \\ \hline 0 \end{array} \qquad \begin{array}{r} 8 \\ -2 \\ \hline 6 \end{array} \qquad \begin{array}{r} 2 \\ -2 \\ \hline 0 \end{array}$$

$$\begin{array}{r} 9 \\ -3 \\ \hline 6 \end{array} \qquad \begin{array}{r} 8 \\ -2 \\ \hline 6 \end{array} \qquad \begin{array}{r} 3 \\ -2 \\ \hline 1 \end{array}$$

Subtraction, Level 1

mathbasics (1,2,3 ... to Pre-Algebra)

Target time: 1 min **Student's time: ____ min**

Solve the following problems:

$$\begin{array}{r} 7 \\ -4 \\ \hline 3 \end{array} \qquad \begin{array}{r} 2 \\ -2 \\ \hline 0 \end{array} \qquad \begin{array}{r} 3 \\ -1 \\ \hline 2 \end{array}$$

$$\begin{array}{r} 7 \\ -5 \\ \hline 2 \end{array} \qquad \begin{array}{r} 3 \\ -2 \\ \hline 1 \end{array} \qquad \begin{array}{r} 3 \\ -2 \\ \hline 1 \end{array}$$

$$\begin{array}{r} 3 \\ -2 \\ \hline 1 \end{array} \qquad \begin{array}{r} 5 \\ -5 \\ \hline 0 \end{array} \qquad \begin{array}{r} 1 \\ -1 \\ \hline 0 \end{array}$$

$$\begin{array}{r} 3 \\ -3 \\ \hline 0 \end{array} \qquad \begin{array}{r} 5 \\ -1 \\ \hline 4 \end{array} \qquad \begin{array}{r} 2 \\ -2 \\ \hline 0 \end{array}$$

$$\begin{array}{r} 7 \\ -2 \\ \hline 5 \end{array} \qquad \begin{array}{r} 9 \\ -5 \\ \hline 4 \end{array} \qquad \begin{array}{r} 4 \\ -1 \\ \hline 3 \end{array}$$

Subtraction, Level 2

Unit: 09
Answers Worksheet: 1

mathbasics (1,2,3 ... to Pre-Algebra) **Target time: 2 min** **Student's time: ____ min**

Solve the following problems:

5814 - 903 4911	5427 - 3901 1526	9112 - 3 9109
9365 - 8182 1183	1474 - 49 1425	7545 - 7365 180
2858 - 1649 1209	8771 - 3191 5580	6352 - 2081 4271
1719 - 913 806	7646 - 7209 437	8115 - 6108 2007
7443 - 272 7171	1587 - 656 931	4679 - 1289 3390

mathbasics (1,2,3 ... to Pre-Algebra) **Target time: 2 min** **Student's time: ____ min**

Solve the following problems:

$$\begin{array}{r} 6646 \\ -716 \\ \hline 5930 \end{array} \qquad \begin{array}{r} 6614 \\ -901 \\ \hline 5713 \end{array} \qquad \begin{array}{r} 1743 \\ -813 \\ \hline 930 \end{array}$$

$$\begin{array}{r} 4665 \\ -3493 \\ \hline 1172 \end{array} \qquad \begin{array}{r} 8175 \\ -814 \\ \hline 7361 \end{array} \qquad \begin{array}{r} 756 \\ -83 \\ \hline 673 \end{array}$$

$$\begin{array}{r} 4086 \\ -2068 \\ \hline 2018 \end{array} \qquad \begin{array}{r} 5085 \\ -3059 \\ \hline 2026 \end{array} \qquad \begin{array}{r} 5281 \\ -5191 \\ \hline 90 \end{array}$$

$$\begin{array}{r} 5561 \\ -4721 \\ \hline 840 \end{array} \qquad \begin{array}{r} 9161 \\ -7116 \\ \hline 2045 \end{array} \qquad \begin{array}{r} 7611 \\ -6711 \\ \hline 900 \end{array}$$

$$\begin{array}{r} 1525 \\ -614 \\ \hline 911 \end{array} \qquad \begin{array}{r} 8236 \\ -5616 \\ \hline 2620 \end{array} \qquad \begin{array}{r} 8434 \\ -2074 \\ \hline 6360 \end{array}$$

Subtraction, Level 2

Unit: 09
Answers Worksheet: 3

mathbasics (1,2,3 ... to Pre-Algebra) **Target time: 2 min Student's time: ____ min**

Solve the following problems:

9927 - 847 9080	2825 - 925 1900	9177 - 4404 4773
7697 - 6678 1019	9977 - 2258 7719	7647 - 6481 1166
7816 - 7571 245	4451 - 1361 3090	3502 - 1902 1600
1128 - 728 400	769 - 483 286	5144 - 5008 136
6546 - 1228 5318	9345 - 9092 253	6392 - 3029 3363

Solve the following problems:

4906 - 2751 = 2155	7076 - 2211 = 4865	4704 - 4521 = 183
5703 - 1173 = 4530	2065 - 1501 = 564	1990 - 1706 = 284
7709 - 2674 = 5035	2036 - 611 = 1425	2101 - 1031 = 1070
1072 - 441 = 631	3105 - 3013 = 92	660 - 403 = 257
8050 - 8041 = 9	7038 - 4831 = 2207	4670 - 1352 = 3318

Subtraction, Level 3

Unit: 10
Answers Worksheet: 2

mathbasics (1,2,3 ... to Pre-Algebra) **Target time: 3 min** **Student's time: ____ min**

Solve the following problems:

3570 - 1149 2421	4850 - 1406 3444	2940 - 13 2927
770 - 714 56	7502 - 3492 4010	5840 - 404 5436
3207 - 3162 45	8905 - 3463 5442	7040 - 4015 3025
2130 - 2105 25	9076 - 3401 5675	2054 - 1921 133
4280 - 4165 115	3013 - 2501 512	6072 - 4821 1251

Solve the following problems:

5103 - 11 5092	1820 - 314 1506	8097 - 6511 1586
7240 - 7035 205	1790 - 1025 765	6260 - 2239 4021
2022 - 201 1821	8906 - 6345 2561	4880 - 228 4652
7403 - 3031 4372	8708 - 4184 4524	9302 - 3232 6070
4049 - 1711 2338	6520 - 4105 2415	3807 - 3036 771

Subtraction, Level 4

Unit: 11
Answers Worksheet: 1

mathbasics (1,2,3 ... to Pre-Algebra) **Target time: 4 min** **Student's time: ____ min**

Solve the following problems:

9051 - 8671 380	4419 - 2596 1823	3502 - 2891 611
4587 - 2796 1791	3960 - 397 3563	9642 - 7566 2076
6426 - 3975 2451	2178 - 1598 580	3848 - 2977 871
955 - 578 377	4404 - 2551 1853	8716 - 4287 4429
8156 - 1068 7088	8671 - 7891 780	8531 - 2861 5670

Solve the following problems:

5551 - 3971 1580	7320 - 2235 5085	4251 - 1881 2370
4321 - 961 3360	2585 - 1397 1188	3716 - 2826 890
5813 - 921 4892	1337 - 444 893	8932 - 2848 6084
5584 - 3396 2188	3404 - 912 2492	5564 - 3672 1892
1755 - 276 1479	3533 - 1296 2237	3860 - 491 3369

Subtraction, Level 4

Unit: 11
Answers Worksheet: 3

mathbasics (1,2,3 ... to Pre-Algebra) **Target time: 4 min** **Student's time: ____ min**

Solve the following problems:

6850 - 2693 4157	9174 - 6692 2482	7544 - 2963 4581
452 - 185 267	3165 - 1076 2089	3755 - 1993 1762
5707 - 2957 2750	9488 - 1099 8389	9565 - 3695 5870
2512 - 356 2156	3705 - 3049 656	1170 - 84 1086
6404 - 2721 3683	6843 - 5375 1468	1626 - 773 853

Multiplication, Level 1

Unit: 12
Answers Worksheet: 1

mathbasics (1,2,3 ... to Pre-Algebra) **Target time: 1 min** **Student's time: ____ min**

Solve the following problems:

$5 \times 7 = 35$	$10 \times 9 = 90$	$8 \times 3 = 24$
$8 \times 8 = 64$	$2 \times 8 = 16$	$7 \times 1 = 7$
$8 \times 12 = 96$	$7 \times 1 = 7$	$7 \times 12 = 84$
$8 \times 6 = 48$	$10 \times 4 = 40$	$3 \times 9 = 27$
$2 \times 2 = 4$	$5 \times 6 = 30$	$4 \times 9 = 36$
$6 \times 9 = 54$	$12 \times 10 = 120$	$9 \times 2 = 18$
$7 \times 10 = 70$	$8 \times 12 = 96$	$10 \times 11 = 110$
$4 \times 3 = 12$	$6 \times 10 = 60$	$9 \times 8 = 72$
$12 \times 12 = 144$	$9 \times 3 = 27$	$5 \times 2 = 10$
$3 \times 4 = 12$	$7 \times 3 = 21$	$4 \times 5 = 20$
$10 \times 7 = 70$	$12 \times 2 = 24$	$3 \times 5 = 15$
$12 \times 6 = 72$	$12 \times 2 = 24$	$8 \times 10 = 80$

Multiplication, Level 1

Unit: 12
Answers Worksheet: 2

mathbasics (1,2,3 ... to Pre-Algebra) **Target time: 1 min** **Student's time: ____ min**

Solve the following problems:

$4 \times 9 = 36$	$7 \times 3 = 21$	$3 \times 5 = 15$
$12 \times 1 = 12$	$9 \times 3 = 27$	$8 \times 11 = 88$
$5 \times 11 = 55$	$9 \times 8 = 72$	$1 \times 4 = 4$
$7 \times 7 = 49$	$9 \times 11 = 99$	$9 \times 4 = 36$
$7 \times 12 = 84$	$4 \times 10 = 40$	$10 \times 7 = 70$
$9 \times 7 = 63$	$2 \times 12 = 24$	$12 \times 2 = 24$
$8 \times 11 = 88$	$7 \times 1 = 7$	$3 \times 5 = 15$
$11 \times 7 = 77$	$9 \times 10 = 90$	$11 \times 3 = 33$
$1 \times 1 = 1$	$9 \times 1 = 9$	$9 \times 9 = 81$
$10 \times 9 = 90$	$5 \times 6 = 30$	$6 \times 1 = 6$
$12 \times 5 = 60$	$1 \times 10 = 10$	$8 \times 4 = 32$
$2 \times 10 = 20$	$10 \times 9 = 90$	$6 \times 1 = 6$

Multiplication, Level 1

mathbasics (1,2,3 ... to Pre-Algebra)

Target time: 1 min   Student's time: ____ min

Solve the following problems:

$11 \times 4 = 44$	$4 \times 3 = 12$	$5 \times 3 = 15$
$11 \times 8 = 88$	$12 \times 5 = 60$	$7 \times 12 = 84$
$11 \times 7 = 77$	$5 \times 12 = 60$	$1 \times 6 = 6$
$1 \times 4 = 4$	$4 \times 9 = 36$	$9 \times 2 = 18$
$4 \times 7 = 28$	$10 \times 6 = 60$	$6 \times 2 = 12$
$12 \times 7 = 84$	$3 \times 9 = 27$	$6 \times 11 = 66$
$2 \times 5 = 10$	$11 \times 4 = 44$	$2 \times 2 = 4$
$8 \times 4 = 32$	$1 \times 8 = 8$	$8 \times 6 = 48$
$1 \times 11 = 11$	$8 \times 8 = 64$	$9 \times 12 = 108$
$12 \times 1 = 12$	$5 \times 7 = 35$	$6 \times 12 = 72$
$8 \times 5 = 40$	$2 \times 3 = 6$	$6 \times 5 = 30$
$12 \times 8 = 96$	$12 \times 7 = 84$	$12 \times 10 = 120$

Multiplication, Level 2

Unit: 13
Answers Worksheet: 1

mathbasics (1,2,3 ... to Pre-Algebra)

Target time: 1 min Student's time: ____ min

Solve the following problems:

$$\begin{array}{r} 5 \\ \times 1 \\ \hline 5 \end{array} \qquad \begin{array}{r} 6 \\ \times 9 \\ \hline 54 \end{array} \qquad \begin{array}{r} 4 \\ \times 6 \\ \hline 24 \end{array}$$

$$\begin{array}{r} 5 \\ \times 5 \\ \hline 25 \end{array} \qquad \begin{array}{r} 8 \\ \times 3 \\ \hline 24 \end{array} \qquad \begin{array}{r} 7 \\ \times 9 \\ \hline 63 \end{array}$$

$$\begin{array}{r} 8 \\ \times 7 \\ \hline 56 \end{array} \qquad \begin{array}{r} 9 \\ \times 5 \\ \hline 45 \end{array} \qquad \begin{array}{r} 2 \\ \times 8 \\ \hline 16 \end{array}$$

$$\begin{array}{r} 4 \\ \times 5 \\ \hline 20 \end{array} \qquad \begin{array}{r} 5 \\ \times 5 \\ \hline 25 \end{array} \qquad \begin{array}{r} 4 \\ \times 5 \\ \hline 20 \end{array}$$

$$\begin{array}{r} 5 \\ \times 3 \\ \hline 15 \end{array} \qquad \begin{array}{r} 2 \\ \times 2 \\ \hline 4 \end{array} \qquad \begin{array}{r} 7 \\ \times 2 \\ \hline 14 \end{array}$$

Multiplication, Level 2

Unit: 13
Answers Worksheet: 2

mathbasics (1,2,3 ... to Pre-Algebra) **Target time: 1 min** **Student's time: ____ min**

Solve the following problems:

$$\begin{array}{r} 7 \\ \times 2 \\ \hline 14 \end{array} \qquad \begin{array}{r} 2 \\ \times 3 \\ \hline 6 \end{array} \qquad \begin{array}{r} 2 \\ \times 6 \\ \hline 12 \end{array}$$

$$\begin{array}{r} 9 \\ \times 7 \\ \hline 63 \end{array} \qquad \begin{array}{r} 2 \\ \times 7 \\ \hline 14 \end{array} \qquad \begin{array}{r} 9 \\ \times 8 \\ \hline 72 \end{array}$$

$$\begin{array}{r} 7 \\ \times 8 \\ \hline 56 \end{array} \qquad \begin{array}{r} 6 \\ \times 6 \\ \hline 36 \end{array} \qquad \begin{array}{r} 3 \\ \times 7 \\ \hline 21 \end{array}$$

$$\begin{array}{r} 5 \\ \times 6 \\ \hline 30 \end{array} \qquad \begin{array}{r} 3 \\ \times 1 \\ \hline 3 \end{array} \qquad \begin{array}{r} 2 \\ \times 6 \\ \hline 12 \end{array}$$

$$\begin{array}{r} 5 \\ \times 6 \\ \hline 30 \end{array} \qquad \begin{array}{r} 9 \\ \times 7 \\ \hline 63 \end{array} \qquad \begin{array}{r} 7 \\ \times 6 \\ \hline 42 \end{array}$$

Multiplication, Level 2

mathbasics (1,2,3 ... to Pre-Algebra)   **Target time: 1 min   Student's time: ____ min**

Solve the following problems:

$$\begin{array}{r} 8 \\ \times 4 \\ \hline 32 \end{array} \qquad \begin{array}{r} 8 \\ \times 9 \\ \hline 72 \end{array} \qquad \begin{array}{r} 7 \\ \times 1 \\ \hline 7 \end{array}$$

$$\begin{array}{r} 6 \\ \times 6 \\ \hline 36 \end{array} \qquad \begin{array}{r} 7 \\ \times 8 \\ \hline 56 \end{array} \qquad \begin{array}{r} 4 \\ \times 7 \\ \hline 28 \end{array}$$

$$\begin{array}{r} 6 \\ \times 1 \\ \hline 6 \end{array} \qquad \begin{array}{r} 5 \\ \times 3 \\ \hline 15 \end{array} \qquad \begin{array}{r} 7 \\ \times 8 \\ \hline 56 \end{array}$$

$$\begin{array}{r} 9 \\ \times 2 \\ \hline 18 \end{array} \qquad \begin{array}{r} 5 \\ \times 3 \\ \hline 15 \end{array} \qquad \begin{array}{r} 2 \\ \times 6 \\ \hline 12 \end{array}$$

$$\begin{array}{r} 8 \\ \times 7 \\ \hline 56 \end{array} \qquad \begin{array}{r} 3 \\ \times 8 \\ \hline 24 \end{array} \qquad \begin{array}{r} 4 \\ \times 1 \\ \hline 4 \end{array}$$

Solve the following problems:

$$\begin{array}{r} 67 \\ \times 3 \\ \hline 201 \end{array} \qquad \begin{array}{r} 59 \\ \times 5 \\ \hline 295 \end{array} \qquad \begin{array}{r} 81 \\ \times 3 \\ \hline 243 \end{array}$$

$$\begin{array}{r} 83 \\ \times 1 \\ \hline 83 \end{array} \qquad \begin{array}{r} 18 \\ \times 5 \\ \hline 90 \end{array} \qquad \begin{array}{r} 82 \\ \times 9 \\ \hline 738 \end{array}$$

$$\begin{array}{r} 15 \\ \times 7 \\ \hline 105 \end{array} \qquad \begin{array}{r} 47 \\ \times 3 \\ \hline 141 \end{array} \qquad \begin{array}{r} 39 \\ \times 5 \\ \hline 195 \end{array}$$

$$\begin{array}{r} 45 \\ \times 7 \\ \hline 315 \end{array} \qquad \begin{array}{r} 85 \\ \times 5 \\ \hline 425 \end{array} \qquad \begin{array}{r} 37 \\ \times 2 \\ \hline 74 \end{array}$$

$$\begin{array}{r} 79 \\ \times 4 \\ \hline 316 \end{array} \qquad \begin{array}{r} 55 \\ \times 5 \\ \hline 275 \end{array} \qquad \begin{array}{r} 74 \\ \times 1 \\ \hline 74 \end{array}$$

Solve the following problems:

$$\begin{array}{r}13\\ \times 4\\ \hline 52\end{array}\qquad\begin{array}{r}37\\ \times 6\\ \hline 222\end{array}\qquad\begin{array}{r}91\\ \times 1\\ \hline 91\end{array}$$

$$\begin{array}{r}91\\ \times 7\\ \hline 637\end{array}\qquad\begin{array}{r}22\\ \times 9\\ \hline 198\end{array}\qquad\begin{array}{r}32\\ \times 9\\ \hline 288\end{array}$$

$$\begin{array}{r}91\\ \times 3\\ \hline 273\end{array}\qquad\begin{array}{r}69\\ \times 3\\ \hline 207\end{array}\qquad\begin{array}{r}63\\ \times 9\\ \hline 567\end{array}$$

$$\begin{array}{r}83\\ \times 9\\ \hline 747\end{array}\qquad\begin{array}{r}68\\ \times 6\\ \hline 408\end{array}\qquad\begin{array}{r}55\\ \times 9\\ \hline 495\end{array}$$

$$\begin{array}{r}83\\ \times 3\\ \hline 249\end{array}\qquad\begin{array}{r}41\\ \times 3\\ \hline 123\end{array}\qquad\begin{array}{r}84\\ \times 1\\ \hline 84\end{array}$$

mathbasics (1,2,3 ... to Pre-Algebra) **Target time: 2 min** **Student's time: ____ min**

Solve the following problems:

$$\begin{array}{r} 40 \\ \times 8 \\ \hline 320 \end{array} \qquad \begin{array}{r} 48 \\ \times 6 \\ \hline 288 \end{array} \qquad \begin{array}{r} 50 \\ \times 3 \\ \hline 150 \end{array}$$

$$\begin{array}{r} 75 \\ \times 4 \\ \hline 300 \end{array} \qquad \begin{array}{r} 14 \\ \times 8 \\ \hline 112 \end{array} \qquad \begin{array}{r} 42 \\ \times 3 \\ \hline 126 \end{array}$$

$$\begin{array}{r} 83 \\ \times 3 \\ \hline 249 \end{array} \qquad \begin{array}{r} 65 \\ \times 5 \\ \hline 325 \end{array} \qquad \begin{array}{r} 63 \\ \times 2 \\ \hline 126 \end{array}$$

$$\begin{array}{r} 50 \\ \times 4 \\ \hline 200 \end{array} \qquad \begin{array}{r} 27 \\ \times 1 \\ \hline 27 \end{array} \qquad \begin{array}{r} 16 \\ \times 7 \\ \hline 112 \end{array}$$

$$\begin{array}{r} 14 \\ \times 8 \\ \hline 112 \end{array} \qquad \begin{array}{r} 98 \\ \times 4 \\ \hline 392 \end{array} \qquad \begin{array}{r} 28 \\ \times 8 \\ \hline 224 \end{array}$$

Solve the following problems:

$\begin{array}{r} 87 \\ \times 33 \\ \hline 2871 \end{array}$	$\begin{array}{r} 77 \\ \times 97 \\ \hline 7469 \end{array}$	$\begin{array}{r} 87 \\ \times 13 \\ \hline 1131 \end{array}$
$\begin{array}{r} 75 \\ \times 97 \\ \hline 7275 \end{array}$	$\begin{array}{r} 19 \\ \times 79 \\ \hline 1501 \end{array}$	$\begin{array}{r} 10 \\ \times 16 \\ \hline 160 \end{array}$
$\begin{array}{r} 94 \\ \times 66 \\ \hline 6204 \end{array}$	$\begin{array}{r} 24 \\ \times 65 \\ \hline 1560 \end{array}$	$\begin{array}{r} 26 \\ \times 46 \\ \hline 1196 \end{array}$
$\begin{array}{r} 54 \\ \times 40 \\ \hline 2160 \end{array}$	$\begin{array}{r} 13 \\ \times 14 \\ \hline 182 \end{array}$	$\begin{array}{r} 25 \\ \times 37 \\ \hline 925 \end{array}$
$\begin{array}{r} 64 \\ \times 56 \\ \hline 3584 \end{array}$	$\begin{array}{r} 59 \\ \times 41 \\ \hline 2419 \end{array}$	$\begin{array}{r} 56 \\ \times 66 \\ \hline 3696 \end{array}$

Solve the following problems:

$$\begin{array}{r} 40 \\ \times 43 \\ \hline 1720 \end{array} \qquad \begin{array}{r} 89 \\ \times 18 \\ \hline 1602 \end{array} \qquad \begin{array}{r} 41 \\ \times 77 \\ \hline 3157 \end{array}$$

$$\begin{array}{r} 21 \\ \times 16 \\ \hline 336 \end{array} \qquad \begin{array}{r} 74 \\ \times 30 \\ \hline 2220 \end{array} \qquad \begin{array}{r} 86 \\ \times 74 \\ \hline 6364 \end{array}$$

$$\begin{array}{r} 36 \\ \times 80 \\ \hline 2880 \end{array} \qquad \begin{array}{r} 40 \\ \times 51 \\ \hline 2040 \end{array} \qquad \begin{array}{r} 46 \\ \times 56 \\ \hline 2576 \end{array}$$

$$\begin{array}{r} 88 \\ \times 91 \\ \hline 8008 \end{array} \qquad \begin{array}{r} 87 \\ \times 91 \\ \hline 7917 \end{array} \qquad \begin{array}{r} 95 \\ \times 13 \\ \hline 1235 \end{array}$$

$$\begin{array}{r} 29 \\ \times 60 \\ \hline 1740 \end{array} \qquad \begin{array}{r} 59 \\ \times 78 \\ \hline 4602 \end{array} \qquad \begin{array}{r} 91 \\ \times 15 \\ \hline 1365 \end{array}$$

Target time: 4 min   Student's time: ____ min

Solve the following problems:

$$\begin{array}{r} 44 \\ \times\,32 \\ \hline 1408 \end{array} \qquad \begin{array}{r} 49 \\ \times\,34 \\ \hline 1666 \end{array} \qquad \begin{array}{r} 40 \\ \times\,80 \\ \hline 3200 \end{array}$$

$$\begin{array}{r} 11 \\ \times\,51 \\ \hline 561 \end{array} \qquad \begin{array}{r} 87 \\ \times\,75 \\ \hline 6525 \end{array} \qquad \begin{array}{r} 72 \\ \times\,73 \\ \hline 5256 \end{array}$$

$$\begin{array}{r} 50 \\ \times\,99 \\ \hline 4950 \end{array} \qquad \begin{array}{r} 54 \\ \times\,81 \\ \hline 4374 \end{array} \qquad \begin{array}{r} 51 \\ \times\,90 \\ \hline 4590 \end{array}$$

$$\begin{array}{r} 38 \\ \times\,40 \\ \hline 1520 \end{array} \qquad \begin{array}{r} 81 \\ \times\,26 \\ \hline 2106 \end{array} \qquad \begin{array}{r} 31 \\ \times\,77 \\ \hline 2387 \end{array}$$

$$\begin{array}{r} 29 \\ \times\,50 \\ \hline 1450 \end{array} \qquad \begin{array}{r} 38 \\ \times\,78 \\ \hline 2964 \end{array} \qquad \begin{array}{r} 29 \\ \times\,29 \\ \hline 841 \end{array}$$

Multiplication, Level 5

Unit: 16
Answers Worksheet: 1

mathbasics (1,2,3 ... to Pre-Algebra)   **Target time: 6 min   Student's time: ____ min**

Solve the following problems:

$$\begin{array}{r} 400 \\ \times 61 \\ \hline 24400 \end{array} \qquad \begin{array}{r} 991 \\ \times 74 \\ \hline 73334 \end{array} \qquad \begin{array}{r} 156 \\ \times 75 \\ \hline 11700 \end{array}$$

$$\begin{array}{r} 905 \\ \times 22 \\ \hline 19910 \end{array} \qquad \begin{array}{r} 797 \\ \times 60 \\ \hline 47820 \end{array} \qquad \begin{array}{r} 864 \\ \times 93 \\ \hline 80352 \end{array}$$

$$\begin{array}{r} 124 \\ \times 83 \\ \hline 10292 \end{array} \qquad \begin{array}{r} 638 \\ \times 50 \\ \hline 31900 \end{array} \qquad \begin{array}{r} 838 \\ \times 78 \\ \hline 65364 \end{array}$$

$$\begin{array}{r} 747 \\ \times 96 \\ \hline 71712 \end{array} \qquad \begin{array}{r} 737 \\ \times 57 \\ \hline 42009 \end{array} \qquad \begin{array}{r} 262 \\ \times 89 \\ \hline 23318 \end{array}$$

$$\begin{array}{r} 622 \\ \times 63 \\ \hline 39186 \end{array} \qquad \begin{array}{r} 510 \\ \times 33 \\ \hline 16830 \end{array} \qquad \begin{array}{r} 316 \\ \times 99 \\ \hline 31284 \end{array}$$

Multiplication, Level 5

Unit: 16
Answers Worksheet: 2

mathbasics (1,2,3 ... to Pre-Algebra) **Target time: 6 min Student's time: ____ min**

Solve the following problems:

$\begin{array}{r} 778 \\ \times 61 \\ \hline 47458 \end{array}$	$\begin{array}{r} 615 \\ \times 76 \\ \hline 46740 \end{array}$	$\begin{array}{r} 360 \\ \times 67 \\ \hline 24120 \end{array}$
$\begin{array}{r} 523 \\ \times 26 \\ \hline 13598 \end{array}$	$\begin{array}{r} 796 \\ \times 32 \\ \hline 25472 \end{array}$	$\begin{array}{r} 769 \\ \times 66 \\ \hline 50754 \end{array}$
$\begin{array}{r} 259 \\ \times 79 \\ \hline 20461 \end{array}$	$\begin{array}{r} 494 \\ \times 79 \\ \hline 39026 \end{array}$	$\begin{array}{r} 295 \\ \times 33 \\ \hline 9735 \end{array}$
$\begin{array}{r} 582 \\ \times 94 \\ \hline 54708 \end{array}$	$\begin{array}{r} 297 \\ \times 31 \\ \hline 9207 \end{array}$	$\begin{array}{r} 517 \\ \times 45 \\ \hline 23265 \end{array}$
$\begin{array}{r} 216 \\ \times 13 \\ \hline 2808 \end{array}$	$\begin{array}{r} 990 \\ \times 62 \\ \hline 61380 \end{array}$	$\begin{array}{r} 376 \\ \times 30 \\ \hline 11280 \end{array}$

Solve the following problems:

$$\begin{array}{r} 623 \\ \times 15 \\ \hline 9345 \end{array} \qquad \begin{array}{r} 823 \\ \times 23 \\ \hline 18929 \end{array} \qquad \begin{array}{r} 824 \\ \times 18 \\ \hline 14832 \end{array}$$

$$\begin{array}{r} 811 \\ \times 34 \\ \hline 27574 \end{array} \qquad \begin{array}{r} 349 \\ \times 60 \\ \hline 20940 \end{array} \qquad \begin{array}{r} 569 \\ \times 11 \\ \hline 6259 \end{array}$$

$$\begin{array}{r} 269 \\ \times 72 \\ \hline 19368 \end{array} \qquad \begin{array}{r} 813 \\ \times 66 \\ \hline 53658 \end{array} \qquad \begin{array}{r} 527 \\ \times 10 \\ \hline 5270 \end{array}$$

$$\begin{array}{r} 896 \\ \times 10 \\ \hline 8960 \end{array} \qquad \begin{array}{r} 951 \\ \times 19 \\ \hline 18069 \end{array} \qquad \begin{array}{r} 328 \\ \times 46 \\ \hline 15088 \end{array}$$

$$\begin{array}{r} 553 \\ \times 44 \\ \hline 24332 \end{array} \qquad \begin{array}{r} 507 \\ \times 54 \\ \hline 27378 \end{array} \qquad \begin{array}{r} 971 \\ \times 78 \\ \hline 75738 \end{array}$$

Multiplication, Level 6

Unit: 17
Answers Worksheet: 1

mathbasics (1,2,3 ... to Pre-Algebra) **Target time: 12 min Student's time: ____ min**

Solve the following problems:

$$\begin{array}{r} 1617 \\ \times\ 832 \\ \hline 1345344 \end{array} \qquad \begin{array}{r} 9298 \\ \times\ 210 \\ \hline 1952580 \end{array} \qquad \begin{array}{r} 2606 \\ \times\ 599 \\ \hline 1560994 \end{array}$$

$$\begin{array}{r} 6459 \\ \times\ 604 \\ \hline 3901236 \end{array} \qquad \begin{array}{r} 6003 \\ \times\ 872 \\ \hline 5234616 \end{array} \qquad \begin{array}{r} 9236 \\ \times\ 495 \\ \hline 4571820 \end{array}$$

$$\begin{array}{r} 7587 \\ \times\ 658 \\ \hline 4992246 \end{array} \qquad \begin{array}{r} 4905 \\ \times\ 679 \\ \hline 3330495 \end{array} \qquad \begin{array}{r} 9244 \\ \times\ 291 \\ \hline 2690004 \end{array}$$

$$\begin{array}{r} 1990 \\ \times\ 644 \\ \hline 1281560 \end{array} \qquad \begin{array}{r} 3797 \\ \times\ 719 \\ \hline 2730043 \end{array} \qquad \begin{array}{r} 7511 \\ \times\ 512 \\ \hline 3845632 \end{array}$$

Solve the following problems:

9347 × 165 = 1542255	2170 × 856 = 1857520	3364 × 554 = 1863656
4735 × 398 = 1884530	3875 × 403 = 1561625	5085 × 548 = 2786580
9031 × 154 = 1390774	1525 × 503 = 767075	9271 × 976 = 9048496
8988 × 685 = 6156780	6345 × 389 = 2468205	3650 × 559 = 2040350

Multiplication, Level 6

mathbasics (1,2,3 ... to Pre-Algebra)

Target time: 12 min Student's time: ____ min

Solve the following problems:

$$\begin{array}{r}5807\\ \times\ 464\\ \hline 2694448\end{array}$$

$$\begin{array}{r}2035\\ \times\ 860\\ \hline 1750100\end{array}$$

$$\begin{array}{r}1837\\ \times\ 854\\ \hline 1568798\end{array}$$

$$\begin{array}{r}3729\\ \times\ 118\\ \hline 440022\end{array}$$

$$\begin{array}{r}9200\\ \times\ 490\\ \hline 4508000\end{array}$$

$$\begin{array}{r}8749\\ \times\ 256\\ \hline 2239744\end{array}$$

$$\begin{array}{r}9449\\ \times\ 348\\ \hline 3288252\end{array}$$

$$\begin{array}{r}5546\\ \times\ 332\\ \hline 1841272\end{array}$$

$$\begin{array}{r}6518\\ \times\ 963\\ \hline 6276834\end{array}$$

$$\begin{array}{r}7807\\ \times\ 554\\ \hline 4325078\end{array}$$

$$\begin{array}{r}1175\\ \times\ 833\\ \hline 978775\end{array}$$

$$\begin{array}{r}9585\\ \times\ 944\\ \hline 9048240\end{array}$$

Solve the following problems:

$8\overline{)5}$ = 0r5	$7\overline{)2}$ = 0r2	$4\overline{)6}$ = 1r2
$3\overline{)7}$ = 2r1	$3\overline{)7}$ = 2r1	$1\overline{)8}$ = 8
$2\overline{)8}$ = 4	$2\overline{)3}$ = 1r1	$5\overline{)6}$ = 1r1
$2\overline{)3}$ = 1r1	$2\overline{)3}$ = 1r1	$2\overline{)4}$ = 2
$5\overline{)6}$ = 1r1	$2\overline{)4}$ = 2	$3\overline{)8}$ = 2r2

Division, Level 1

Unit: 18
Answers Worksheet: 2

mathbasics (1,2,3 ... to Pre-Algebra) **Target time: 2 min** **Student's time: ____ min**

Solve the following problems:

$1\overline{)5}$ = 5	$1\overline{)2}$ = 2	$2\overline{)8}$ = 4
$5\overline{)7}$ = 1r2	$2\overline{)1}$ = 0r1	$1\overline{)7}$ = 7
$1\overline{)2}$ = 2	$6\overline{)7}$ = 1r1	$1\overline{)2}$ = 2
$1\overline{)8}$ = 8	$1\overline{)2}$ = 2	$2\overline{)4}$ = 2
$8\overline{)3}$ = 0r3	$2\overline{)9}$ = 4r1	$9\overline{)2}$ = 0r2

Division, Level 1

Unit: 18
Answers Worksheet: 3

mathbasics (1,2,3 ... to Pre-Algebra) **Target time: 2 min** **Student's time: ____ min**

Solve the following problems:

$8\overline{)9}$ = 1r1	$4\overline{)5}$ = 1r1	$3\overline{)4}$ = 1r1
$2\overline{)3}$ = 1r1	$4\overline{)7}$ = 1r3	$4\overline{)6}$ = 1r2
$5\overline{)6}$ = 1r1	$4\overline{)6}$ = 1r2	$5\overline{)8}$ = 1r3
$6\overline{)9}$ = 1r3	$4\overline{)8}$ = 2	$8\overline{)9}$ = 1r1
$2\overline{)5}$ = 2r1	$9\overline{)5}$ = 0r5	$2\overline{)7}$ = 3r1

Division, Level 2

Unit: 19
Answers Worksheet: 1

mathbasics (1,2,3 ... to Pre-Algebra) **Target time: 2 min Student's time: ____ min**

Solve the following problems:

$4\overline{)923}$ = 230r3

$7\overline{)786}$ = 112r2

$2\overline{)918}$ = 459

$2\overline{)910}$ = 455

$8\overline{)892}$ = 111r4

$5\overline{)852}$ = 170r2

$4\overline{)538}$ = 134r2

$1\overline{)991}$ = 991

$5\overline{)503}$ = 100r3

mathbasics (1,2,3 ... to Pre-Algebra) **Target time: 2 min** **Student's time: ____ min**

Solve the following problems:

$3\overline{)934}$ = 311r1	$1\overline{)964}$ = 964	$8\overline{)916}$ = 114r4
$2\overline{)389}$ = 194r1	$3\overline{)342}$ = 114	$2\overline{)829}$ = 414r1
$5\overline{)698}$ = 139r3	$3\overline{)397}$ = 132r1	$2\overline{)545}$ = 272r1

Solve the following problems:

$4\overline{)582}$ = 145r2

$5\overline{)840}$ = 168

$2\overline{)240}$ = 120

$5\overline{)913}$ = 182r3

$5\overline{)649}$ = 129r4

$2\overline{)729}$ = 364r1

$7\overline{)779}$ = 111r2

$4\overline{)460}$ = 115

$7\overline{)919}$ = 131r2

Division, Level 3

Unit: 20
Answers Worksheet: 1

mathbasics (1,2,3 ... to Pre-Algebra) **Target time: 12 min** **Student's time: ____ min**

Solve the following problems:

$36\overline{)8042}$ = 223r14

$83\overline{)8455}$ = 101r72

$57\overline{)5670}$ = 99r27

$31\overline{)3143}$ = 101r12

$57\overline{)3890}$ = 68r14

$76\overline{)3372}$ = 44r28

$66\overline{)6536}$ = 99r2

$43\overline{)7218}$ = 167r37

$12\overline{)2269}$ = 189r1

Division, Level 3

Unit: 20
Answers Worksheet: 2

mathbasics (1,2,3 ... to Pre-Algebra) **Target time: 12 min Student's time: ____ min**

Solve the following problems:

$95\overline{)5465}$ = 57r50	$24\overline{)7595}$ = 316r11	$47\overline{)3314}$ = 70r24
$95\overline{)9959}$ = 104r79	$17\overline{)2467}$ = 145r2	$64\overline{)8470}$ = 132r22
$53\overline{)9098}$ = 171r35	$65\overline{)3691}$ = 56r51	$75\overline{)2307}$ = 30r57

Solve the following problems:

116r30 83)9658	93r26 44)4118	60r13 35)2113
143r52 64)9204	129r32 76)9836	123r13 64)7885
285r20 21)6005	108r25 33)3589	197r31 36)7123

Average, Median, Mode

Unit: 21
Answers Worksheet: 1

mathbasics (1,2,3 ... to Pre-Algebra) **Target time: 5 min Student's time: ____ min**

I. Calculate the Average (Mean) of the numbers:

80, 60, 80, 16	104, 58, 94, 72, 88, 70	63, 16, 28, 75, 43
59	81	45

II. Calculate the Median of the numbers:

53, 15, 25, 37, 63, 15	86, 71, 24, 1, 33, 80	37, 93, 9, 47, 69, 4
31	52	42

III. Calculate the Mode of the numbers:

57, 73, 29, 29	80, 92, 58, 58	44, 73, 9, 9, 80
29	58	9

Average, Median, Mode

Unit: 21
Answers Worksheet: 2

mathbasics (1,2,3 ... to Pre-Algebra) **Target time: 5 min** **Student's time: ____ min**

I. Calculate the Average (Mean) of the numbers:

101, 11, 6, 80, 22	13, 46, 33, 40	88, 84, 39, 93
44	33	76

II. Calculate the Median of the numbers:

20, 57, 20, 75, 56	95, 7, 10, 68, 31, 15	34, 16, 21, 37, 10, 21
56	23	21

III. Calculate the Mode of the numbers:

93, 79, 45, 72, 45	88, 8, 15, 93, 31	90, 75, 12, 11, 76, 38
45	none	none

Target time: 5 min **Student's time: ____ min**

I. Calculate the Average (Mean) of the numbers:

56, 63, 40, 57	53, 59, 49, 24, 71, 86	66, 96, 14, 5, 66, 35
54	57	47

II. Calculate the Median of the numbers:

92, 12, 89, 71	71, 5, 99, 23, 98, 63	10, 17, 72, 75, 71, 14
80	67	44

III. Calculate the Mode of the numbers:

69, 69, 33, 43	2, 9, 45, 56, 93	92, 14, 70, 62
69	none	none

I.

4^1	7^1	4^5
4	7	1024

II.

What power of 8 is equal to 262144? 6

What power of 7 is equal to 16807? 5

III. Write using exponents:

9 to the first power:	9^1	2 to the sixth power:	2^6
7 cubed:	7^3	2 to the first power:	2^1
4 to the first power:	4^1	9 to the sixth power:	9^6

IV.

$\sqrt[5]{243}$	$\sqrt[6]{64}$	$\sqrt[4]{16}$	$\sqrt[3]{8}$
3	2	2	2

I.

9^1	6^4	6^6
9	1296	46656

II.

What power of 5 is equal to 25? 2

What power of 7 is equal to 1? 0

III. Write using exponents:

3 to the fifth power:	3^5	4 to the first power:	4^1
6 to the sixth power:	6^6	8 to the fourth power:	8^4
3 to the power of zero:	3^0	8 to the first power:	8^1

IV.

$\sqrt[6]{64}$	$\sqrt[1]{9}$	$\sqrt[1]{9}$	$\sqrt[6]{117649}$
2	9	9	7

I.

4^4	3^6	4^5
256	729	1024

II.

What power of 8 is equal to 32768? 5

What power of 3 is equal to 27? 3

III. Write using exponents:

8 squared:	8^2	3 to the power of zero:	3^0
6 to the fifth power:	6^5	2 squared:	2^2
5 to the first power:	5^1	6 to the first power:	6^1

IV.

$\sqrt[4]{256}$	$\sqrt{9}$	$\sqrt[1]{7}$	$\sqrt[3]{64}$
4	3	7	4

Fractions, Simple to Mixed

Unit: 23
Answers Worksheet: 1

mathbasics (1,2,3 ... to Pre-Algebra) **Target time: 12 min Student's time: ____ min**

Convert the fraction to a mixed number:

$\frac{684}{12} = 57$

$\frac{589}{83} = 7\frac{8}{83}$

$\frac{159}{52} = 3\frac{3}{52}$

$\frac{614}{38} = 16\frac{6}{38}$

$\frac{579}{15} = 38\frac{9}{15}$

$\frac{182}{66} = 2\frac{50}{66}$

$\frac{352}{21} = 16\frac{16}{21}$

$\frac{430}{67} = 6\frac{28}{67}$

$\frac{443}{81} = 5\frac{38}{81}$

$\frac{762}{53} = 14\frac{20}{53}$

$\frac{57}{44} = 1\frac{13}{44}$

$\frac{594}{67} = 8\frac{58}{67}$

Fractions, Simple to Mixed

Unit: 23
Answers Worksheet: 2

mathbasics (1,2,3 ... to Pre-Algebra) **Target time: 12 min Student's time: ____ min**

Convert the fraction to a mixed number:

$\frac{496}{34} = 14\frac{20}{34}$ $\frac{358}{29} = 12\frac{10}{29}$ $\frac{340}{73} = 4\frac{48}{73}$

$\frac{781}{63} = 12\frac{25}{63}$ $\frac{790}{12} = 65\frac{10}{12}$ $\frac{417}{26} = 16\frac{1}{26}$

$\frac{360}{27} = 13\frac{9}{27}$ $\frac{372}{59} = 6\frac{18}{59}$ $\frac{710}{47} = 15\frac{5}{47}$

$\frac{358}{76} = 4\frac{54}{76}$ $\frac{842}{20} = 42\frac{2}{20}$ $\frac{932}{83} = 11\frac{19}{83}$

Convert the fraction to a mixed number:

$\frac{461}{26} = 17\frac{19}{26}$ $\frac{376}{77} = 4\frac{68}{77}$ $\frac{445}{64} = 6\frac{61}{64}$

$\frac{574}{78} = 7\frac{28}{78}$ $\frac{308}{19} = 16\frac{4}{19}$ $\frac{794}{100} = 7\frac{94}{100}$

$\frac{935}{39} = 23\frac{38}{39}$ $\frac{247}{47} = 5\frac{12}{47}$ $\frac{663}{46} = 14\frac{19}{46}$

$\frac{963}{62} = 15\frac{33}{62}$ $\frac{890}{85} = 10\frac{40}{85}$ $\frac{568}{71} = 8$

Fractions, Mixed to Simple

Unit: 24
Answers Worksheet: 1

mathbasics (1,2,3 ... to Pre-Algebra) **Target time: 6 min Student's time: ____ min**

Convert the mixed number to a simple fraction:

$3\frac{27}{607} = \frac{1848}{607}$

$8\frac{42}{690} = \frac{5562}{690}$

$20\frac{98}{632} = \frac{12738}{632}$

$15\frac{3}{863} = \frac{12948}{863}$

$8\frac{71}{488} = \frac{3975}{488}$

$16\frac{36}{366} = \frac{5892}{366}$

$7\frac{60}{586} = \frac{4162}{586}$

$1\frac{7}{326} = \frac{333}{326}$

$12\frac{35}{243} = \frac{2951}{243}$

$10\frac{29}{978} = \frac{9809}{978}$

$11\frac{61}{994} = \frac{10995}{994}$

$9\frac{5}{572} = \frac{5153}{572}$

Fractions, Mixed to Simple

mathbasics (1,2,3 ... to Pre-Algebra)   **Target time: 6 min   Student's time: ____ min**

Convert the mixed number to a simple fraction:

$13\frac{31}{941} = \frac{12264}{941}$ $17\frac{82}{732} = \frac{12526}{732}$ $2\frac{71}{503} = \frac{1077}{503}$

$14\frac{85}{693} = \frac{9787}{693}$ $16\frac{37}{779} = \frac{12501}{779}$ $17\frac{8}{468} = \frac{7964}{468}$

$19\frac{89}{222} = \frac{4307}{222}$ $20\frac{15}{161} = \frac{3235}{161}$ $13\frac{6}{571} = \frac{7429}{571}$

$14\frac{63}{300} = \frac{4263}{300}$ $15\frac{17}{781} = \frac{11732}{781}$ $19\frac{96}{734} = \frac{14042}{734}$

Fractions, Mixed to Simple

Unit: 24
Answers Worksheet: 3

Target time: 6 min   Student's time: ____ min

Convert the mixed number to a simple fraction:

$6\frac{82}{122} = \frac{814}{122}$ $8\frac{34}{460} = \frac{3714}{460}$ $20\frac{17}{783} = \frac{15677}{783}$

$19\frac{40}{504} = \frac{9616}{504}$ $3\frac{35}{148} = \frac{479}{148}$ $18\frac{82}{777} = \frac{14068}{777}$

$10\frac{97}{129} = \frac{1387}{129}$ $14\frac{9}{645} = \frac{9039}{645}$ $2\frac{32}{314} = \frac{660}{314}$

$15\frac{18}{562} = \frac{8448}{562}$ $12\frac{70}{149} = \frac{1858}{149}$ $1\frac{49}{765} = \frac{814}{765}$

Prime Numbers, Prime Factors

Unit: 25
Answers Worksheet: 1

mathbasics (1,2,3 ... to Pre-Algebra) **Target time: 19 min Student's time: ____ min**

Find the prime factors, arrange in order, and express as exponent:

$60 = 2^2 \times 3 \times 5$

$108 = 2^2 \times 3^3$

$5670 = 2 \times 3^4 \times 5 \times 7$

$214375 = 5^4 \times 7^3$

$70 = 2 \times 5 \times 7$

$280 = 2^3 \times 5 \times 7$

$735 = 3 \times 5 \times 7^2$

$11025 = 3^2 \times 5^2 \times 7^2$

$17640 = 2^3 \times 3^2 \times 5 \times 7^2$

$2625 = 3 \times 5^3 \times 7$

$1050 = 2 \times 3 \times 5^2 \times 7$

$34300 = 2^2 \times 5^2 \times 7^3$

Prime Numbers, Prime Factors

mathbasics (1,2,3 ... to Pre-Algebra) **Target time: 19 min** **Student's time: ____ min**

Find the prime factors, arrange in order, and express as exponent:

$7000 = 2^3 \times 5^3 \times 7$

$7000 = 2^3 \times 5^3 \times 7$

$5145 = 3 \times 5 \times 7^3$

$441 = 3^2 \times 7^2$

$50 = 2 \times 5^2$

$36750 = 2 \times 3 \times 5^3 \times 7^2$

$1890 = 2 \times 3^3 \times 5 \times 7$

$12600 = 2^3 \times 3^2 \times 5^2 \times 7$

$315 = 3^2 \times 5 \times 7$

$30 = 2 \times 3 \times 5$

$60 = 2^2 \times 3 \times 5$

$2835 = 3^4 \times 5 \times 7$

Target time: 19 min **Student's time: ____ min**

Find the prime factors, arrange in order, and express as exponent:

$7350 = 2 \times 3 \times 5^2 \times 7^2$

$150 = 2 \times 3 \times 5^2$

$1323 = 3^3 \times 7^2$

$23814 = 2 \times 3^5 \times 7^2$

$120 = 2^3 \times 3 \times 5$

$90 = 2 \times 3^2 \times 5$

$30 = 2 \times 3 \times 5$

$5250 = 2 \times 3 \times 5^3 \times 7$

$735 = 3 \times 5 \times 7^2$

$3780 = 2^2 \times 3^3 \times 5 \times 7$

$3780 = 2^2 \times 3^3 \times 5 \times 7$

$138915 = 3^4 \times 5 \times 7^3$

Target time: 36 min Student's time: ____ min

Find the GCD:

240 and 882	630 and 21875	56250 and 20580
$6 = 1\times2\times3$	$35 = 1\times5\times7$	$30 = 1\times2\times3\times5$
4725 and 6174	6300 and 51450	60 and 154350
$63 = 1\times3\times3\times7$	$1050 = 1\times2\times3\times5\times5\times7$	$30 = 1\times2\times3\times5$
300 and 100	8 and 1125	1715 and 48020
$100 = 1\times2\times2\times5\times5$	$1 = 1$	$1715 = 1\times5\times7\times7\times7$
34300 and 315	1890 and 18	18 and 2058
$35 = 1\times5\times7$	$18 = 1\times2\times3\times3$	$6 = 1\times2\times3$
350 and 6300	13230 and 12348	24010 and 73500
$350 = 1\times2\times5\times5\times7$	$882 = 1\times2\times3\times3\times7\times7$	$490 = 1\times2\times5\times7\times7$
392 and 92610	20580 and 13230	216090 and 588
$98 = 1\times2\times7\times7$	$1470 = 1\times2\times3\times5\times7\times7$	$294 = 1\times2\times3\times7\times7$

Greatest Common Divisor

Unit: 26
Answers Worksheet: 2

mathbasics (1,2,3 ... to Pre-Algebra) **Target time: 36 min Student's time: ____ min**

Find the GCD:

90 and 46305	1715 and 490	3600 and 51450
45 = 1×3×3×5	245 = 1×5×7×7	150 = 1×2×3×5×5
7350 and 28	210 and 18900	26250 and 73500
14 = 1×2×7	210 = 1×2×3×5×7	5250 = 1×2×3×5×5×5×7
17640 and 105	3528 and 18	29400 and 16
105 = 1×3×5×7	18 = 1×2×3×3	8 = 1×2×2×2
125 and 14700	1960 and 882	68600 and 44100
25 = 1×5×5	98 = 1×2×7×7	4900 = 1×2×2×5×5×7×7
63 and 750	280 and 490	4375 and 630
3 = 1×3	70 = 1×2×5×7	35 = 1×5×7
735 and 72030	375 and 49000	675 and 17640
735 = 1×3×5×7×7	125 = 1×5×5×5	45 = 1×3×3×5

Greatest Common Divisor

Unit: 26
Answers Worksheet: 3

mathbasics (1,2,3 ... to Pre-Algebra) **Target time: 36 min** **Student's time: ____ min**

Find the GCD:

4050 and 257250	36750 and 180	150 and 36
150 = 1×2×3×5×5	30 = 1×2×3×5	6 = 1×2×3
4900 and 375	90 and 90	10290 and 14700
25 = 1×5×5	90 = 1×2×3×3×5	1470 = 1×2×3×5×7×7
1372 and 20	1050 and 75	28 and 420
4 = 1×2×2	75 = 1×3×5×5	28 = 1×2×2×7
120 and 125	504 and 70	189 and 1029
5 = 1×5	14 = 1×2×7	21 = 1×3×7
2100 and 61740	168 and 1620	882 and 4725
420 = 1×2×2×3×5×7	12 = 1×2×2×3	63 = 1×3×3×7
1176 and 50	4200 and 1176490	245 and 61740
2 = 1×2	70 = 1×2×5×7	245 = 1×5×7×7

Reducing Fractions

Unit: 27
Answers Worksheet: 1

mathbasics (1,2,3 ... to Pre-Algebra)   **Target time: 13 min   Student's time: ____ min**

Reduce each fraction to its lowest terms:

$\frac{75}{375} = \frac{1}{5}$

$\frac{50}{200} = \frac{1}{4}$

$\frac{210}{350} = \frac{3}{5}$

$\frac{343}{686} = \frac{1}{2}$

$\frac{70}{140} = \frac{1}{2}$

$\frac{30}{240} = \frac{1}{8}$

$\frac{7}{8}$ in terms of 16ths

$= \frac{14}{16}$

$\frac{3}{5}$ in terms of 15ths

$= \frac{9}{15}$

$\frac{1}{2}$ in terms of 22nds

$= \frac{11}{22}$

$\frac{5}{8}$ in terms of 104ths

$= \frac{65}{104}$

$\frac{5}{8}$ in terms of 88ths

$= \frac{55}{88}$

$\frac{7}{8}$ in terms of 40ths

$= \frac{35}{40}$

Reducing Fractions

mathbasics (1,2,3 ... to Pre-Algebra) Target time: 13 min Student's time: ____ min

Reduce each fraction to its lowest terms:

$\frac{105}{210} = \frac{1}{2}$

$\frac{36}{96} = \frac{3}{8}$

$\frac{350}{560} = \frac{5}{8}$

$\frac{315}{525} = \frac{3}{5}$

$\frac{28}{140} = \frac{1}{5}$

$\frac{112}{140} = \frac{4}{5}$

$\frac{5}{8}$ in terms of 56ths $= \frac{35}{56}$

$\frac{3}{8}$ in terms of 40ths $= \frac{15}{40}$

$\frac{7}{8}$ in terms of 16ths $= \frac{14}{16}$

$\frac{3}{5}$ in terms of 25ths $= \frac{15}{25}$

$\frac{3}{5}$ in terms of 35ths $= \frac{21}{35}$

$\frac{1}{2}$ in terms of 26ths $= \frac{13}{26}$

Reducing Fractions

Unit: 27
Answers Worksheet: 3

mathbasics (1,2,3 ... to Pre-Algebra) **Target time: 13 min** **Student's time: ____ min**

Reduce each fraction to its lowest terms:

$\frac{225}{600} = \frac{3}{8}$

$\frac{98}{196} = \frac{1}{2}$

$\frac{40}{100} = \frac{2}{5}$

$\frac{125}{500} = \frac{1}{4}$

$\frac{90}{144} = \frac{5}{8}$

$\frac{81}{108} = \frac{3}{4}$

$\frac{5}{8}$ in terms of 104ths

$= \frac{65}{104}$

$\frac{1}{8}$ in terms of 104ths

$= \frac{13}{104}$

$\frac{2}{5}$ in terms of 35ths

$= \frac{14}{35}$

$\frac{5}{8}$ in terms of 104ths

$= \frac{65}{104}$

$\frac{5}{8}$ in terms of 16ths

$= \frac{10}{16}$

$\frac{1}{5}$ in terms of 65ths

$= \frac{13}{65}$

Fractions, Addition

Unit: 28
Answers Worksheet: 1

mathbasics (1,2,3 ... to Pre-Algebra) **Target time: 11 min Student's time: ____ min**

Solve the following problems (leave the answer in reduced, improper fraction form):

$\frac{18}{87} + \frac{23}{87} = \frac{41}{87}$

$\frac{19}{79} + \frac{37}{79} = \frac{56}{79}$

$\frac{4}{83} + \frac{39}{83} = \frac{43}{83}$

$\frac{2}{3} + \frac{16}{2} = \frac{26}{3}$

$\frac{38}{3} + \frac{31}{5} = \frac{283}{15}$

$\frac{12}{5} + \frac{23}{3} = \frac{151}{15}$

$\frac{10}{11} + \frac{19}{3} = \frac{239}{33}$

$\frac{2}{3} + \frac{30}{5} = \frac{20}{3}$

$\frac{8}{11} + \frac{39}{2} = \frac{445}{22}$

$8\frac{27}{3} + 7\frac{19}{11} = \frac{283}{11}$

$2\frac{21}{7} + 7\frac{19}{11} = \frac{151}{11}$

$2\frac{39}{11} + 8\frac{17}{3} = \frac{634}{33}$

Fractions, Addition

Unit: 28
Answers Worksheet: 2

mathbasics (1,2,3 ... to Pre-Algebra) **Target time: 11 min Student's time: ____ min**

Solve the following problems (leave the answer in reduced, improper fraction form):

$\frac{22}{78}+\frac{39}{78}=\frac{61}{78}$

$\frac{26}{66}+\frac{35}{66}=\frac{61}{66}$

$\frac{34}{77}+\frac{20}{77}=\frac{54}{77}$

$\frac{20}{11}+\frac{38}{3}=\frac{478}{33}$

$\frac{14}{11}+\frac{5}{5}=\frac{25}{11}$

$\frac{35}{11}+\frac{21}{5}=\frac{406}{55}$

$\frac{24}{7}+\frac{1}{5}=\frac{127}{35}$

$\frac{1}{5}+\frac{4}{7}=\frac{27}{35}$

$\frac{23}{5}+\frac{28}{2}=\frac{93}{5}$

$1\frac{25}{7}+4\frac{29}{5}=\frac{503}{35}$

$4\frac{2}{7}+7\frac{8}{11}=\frac{925}{77}$

$5\frac{39}{3}+8\frac{16}{5}=\frac{146}{5}$

Solve the following problems (leave the answer in reduced, improper fraction form):

$\frac{22}{80} + \frac{1}{80} = \frac{23}{80}$

$\frac{24}{91} + \frac{31}{91} = \frac{55}{91}$

$\frac{23}{79} + \frac{29}{79} = \frac{52}{79}$

$\frac{9}{5} + \frac{22}{7} = \frac{173}{35}$

$\frac{25}{2} + \frac{19}{5} = \frac{163}{10}$

$\frac{18}{11} + \frac{7}{7} = \frac{29}{11}$

$\frac{35}{3} + \frac{20}{7} = \frac{305}{21}$

$\frac{39}{3} + \frac{30}{7} = \frac{121}{7}$

$\frac{13}{3} + \frac{25}{7} = \frac{166}{21}$

$1\frac{36}{5} + 4\frac{37}{3} = \frac{368}{15}$

$4\frac{36}{2} + 6\frac{16}{3} = \frac{100}{3}$

$2\frac{29}{7} + 9\frac{19}{2} = \frac{345}{14}$

Fractions, Subtraction

mathbasics (1,2,3 ... to Pre-Algebra) Target time: 10 min Student's time: ____ min

Solve the following problems
(leave the answer in reduced, improper fraction form):

$\frac{36}{83} - \frac{13}{83} = \frac{23}{83}$

$\frac{33}{76} - \frac{12}{76} = \frac{21}{76}$

$\frac{16}{51} - \frac{9}{51} = \frac{7}{51}$

$\frac{11}{6} - \frac{17}{15} = \frac{7}{10}$

$\frac{15}{5} - \frac{9}{6} = \frac{3}{2}$

$\frac{4}{4} - \frac{8}{14} = \frac{3}{7}$

$\frac{7}{6} - \frac{3}{3} = \frac{1}{6}$

$\frac{4}{2} - \frac{9}{13} = \frac{17}{13}$

$\frac{13}{3} - \frac{15}{6} = \frac{11}{6}$

$8\frac{4}{2} - 2\frac{10}{8} = \frac{27}{4}$

$9\frac{1}{9} - 1\frac{6}{3} = \frac{55}{9}$

$6\frac{5}{5} - 2\frac{6}{2} = \frac{2}{1}$

Fractions, Subtraction

Unit: 29
Answers Worksheet: 2

mathbasics (1,2,3 ... to Pre-Algebra) **Target time: 10 min** **Student's time: ____ min**

Solve the following problems
(leave the answer in reduced, improper fraction form):

$\frac{28}{93} - \frac{17}{93} = \frac{11}{93}$

$\frac{19}{51} - \frac{16}{51} = \frac{1}{17}$

$\frac{39}{80} - \frac{1}{80} = \frac{19}{40}$

$\frac{12}{3} - \frac{13}{13} = \frac{3}{1}$

$\frac{15}{6} - \frac{1}{14} = \frac{17}{7}$

$\frac{12}{4} - \frac{2}{4} = \frac{5}{2}$

$\frac{9}{2} - \frac{19}{7} = \frac{25}{14}$

$\frac{13}{6} - \frac{5}{12} = \frac{7}{4}$

$\frac{7}{5} - \frac{10}{10} = \frac{2}{5}$

$11\frac{10}{3} - 2\frac{11}{4} = \frac{115}{12}$

$12\frac{6}{8} - 3\frac{4}{9} = \frac{335}{36}$

$11\frac{5}{7} - 4\frac{5}{5} = \frac{47}{7}$

Fractions, Subtraction

Unit: 29
Answers Worksheet: 3

mathbasics (1,2,3 ... to Pre-Algebra) **Target time: 10 min Student's time: ____ min**

Solve the following problems (leave the answer in reduced, improper fraction form):

$\frac{33}{88} - \frac{4}{88} = \frac{29}{88}$

$\frac{35}{72} - \frac{8}{72} = \frac{3}{8}$

$\frac{23}{78} - \frac{5}{78} = \frac{3}{13}$

$\frac{9}{3} - \frac{11}{11} = \frac{2}{1}$

$\frac{17}{5} - \frac{6}{3} = \frac{7}{5}$

$\frac{4}{5} - \frac{2}{12} = \frac{19}{30}$

$\frac{2}{3} - \frac{5}{9} = \frac{1}{9}$

$\frac{13}{6} - \frac{1}{4} = \frac{23}{12}$

$\frac{16}{3} - \frac{13}{12} = \frac{17}{4}$

$15\frac{2}{10} - 4\frac{10}{11} = \frac{566}{55}$

$10\frac{4}{3} - 3\frac{10}{9} = \frac{65}{9}$

$10\frac{10}{10} - 3\frac{3}{8} = \frac{61}{8}$

Least Common Denominator

mathbasics (1,2,3 ... to Pre-Algebra) **Target time: 25 min Student's time: ____ min**

Find the LCD and Solve the following problems (leave the answer in improper fraction form):

$\frac{8}{21} + \frac{7}{42} = \frac{23}{42}$

LCD of 21 and 42 = 42

$\frac{3}{98} + \frac{4}{35} = \frac{71}{490}$

LCD of 98 and 35 = 490

$\frac{3}{25} + \frac{7}{21} = \frac{34}{75}$

LCD of 25 and 21 = 525

$\frac{6}{21} - \frac{1}{45} = \frac{83}{315}$

LCD of 21 and 45 = 315

$\frac{9}{42} - \frac{7}{105} = \frac{31}{210}$

LCD of 42 and 105 = 210

$\frac{3}{70} + \frac{8}{35} = \frac{19}{70}$

LCD of 70 and 35 = 70

$4\frac{5}{18} - 3\frac{7}{42} = \frac{10}{9}$

LCD of 18 and 42 = 126

$2\frac{8}{15} + 3\frac{7}{45} = \frac{256}{45}$

LCD of 15 and 45 = 45

$1\frac{1}{49} + 1\frac{1}{12} = \frac{1237}{588}$

LCD of 49 and 12 = 588

$4\frac{4}{14} + 1\frac{4}{15} = \frac{583}{105}$

LCD of 14 and 15 = 210

$2\frac{9}{30} + 5\frac{1}{343} = \frac{25049}{3430}$

LCD of 30 and 343 = 10290

$4\frac{4}{10} + 1\frac{1}{105} = \frac{568}{105}$

LCD of 10 and 105 = 210

Find the LCD and Solve the following problems (leave the answer in improper fraction form):

$\frac{8}{21} - \frac{5}{70} = \frac{13}{42}$

LCD of 21 and 70 = 210

$\frac{2}{15} - \frac{2}{70} = \frac{11}{105}$

LCD of 15 and 70 = 210

$\frac{8}{18} - \frac{5}{35} = \frac{19}{63}$

LCD of 18 and 35 = 630

$\frac{8}{30} + \frac{2}{35} = \frac{34}{105}$

LCD of 30 and 35 = 210

$\frac{8}{21} + \frac{3}{45} = \frac{47}{105}$

LCD of 21 and 45 = 315

$\frac{3}{98} + \frac{2}{20} = \frac{32}{245}$

LCD of 98 and 20 = 980

$2\frac{2}{6} - 1\frac{4}{10} = \frac{14}{15}$

LCD of 6 and 10 = 30

$1\frac{6}{14} + 5\frac{2}{105} = \frac{677}{105}$

LCD of 14 and 105 = 210

$5\frac{3}{12} - 2\frac{1}{147} = \frac{1907}{588}$

LCD of 12 and 147 = 588

$2\frac{9}{10} - 2\frac{7}{30} = \frac{2}{3}$

LCD of 10 and 30 = 30

$4\frac{9}{70} - 3\frac{5}{10} = \frac{22}{35}$

LCD of 70 and 10 = 70

$5\frac{2}{35} - 3\frac{4}{9} = \frac{508}{315}$

LCD of 35 and 9 = 315

Least Common Denominator

Unit: 30
Answers Worksheet: 3

mathbasics (1,2,3 ... to Pre-Algebra) **Target time: 25 min Student's time: ____ min**

Find the LCD and Solve the following problems (leave the answer in improper fraction form):

$\frac{1}{98} - \frac{3}{343} = \frac{1}{686}$

LCD of 98 and 343 = 686

$\frac{7}{4} - \frac{2}{21} = \frac{139}{84}$

LCD of 4 and 21 = 84

$\frac{5}{21} - \frac{6}{70} = \frac{16}{105}$

LCD of 21 and 70 = 210

$\frac{9}{8} + \frac{2}{35} = \frac{331}{280}$

LCD of 8 and 35 = 280

$\frac{4}{21} - \frac{3}{30} = \frac{19}{210}$

LCD of 21 and 30 = 210

$\frac{9}{35} - \frac{3}{28} = \frac{3}{20}$

LCD of 35 and 28 = 140

$5\frac{7}{21} + 3\frac{5}{35} = \frac{178}{21}$

LCD of 21 and 35 = 105

$4\frac{8}{15} - 2\frac{1}{21} = \frac{87}{35}$

LCD of 15 and 21 = 105

$2\frac{7}{21} - 2\frac{3}{63} = \frac{2}{7}$

LCD of 21 and 63 = 63

$3\frac{5}{15} - 1\frac{4}{10} = \frac{29}{15}$

LCD of 15 and 10 = 30

$2\frac{7}{45} + 1\frac{2}{70} = \frac{1003}{315}$

LCD of 45 and 70 = 630

$5\frac{9}{98} - 3\frac{1}{35} = \frac{1011}{490}$

LCD of 98 and 35 = 490

Fractions, Multiplication

Unit: 31
Answers Worksheet: 1

mathbasics (1,2,3 ... to Pre-Algebra) **Target time: 3 min Student's time: ____ min**

Multiply the fractions and reduce to simplest fraction:

$\frac{10}{10} \times \frac{5}{6} = \frac{5}{6}$

$\frac{7}{8} \times \frac{8}{6} = \frac{7}{6}$

$\frac{2}{7} \times \frac{3}{2} = \frac{3}{7}$

$\frac{10}{3} \times \frac{8}{5} = \frac{16}{3}$

$\frac{2}{4} \times \frac{7}{7} = \frac{1}{2}$

$\frac{6}{2} \times \frac{7}{5} = \frac{21}{5}$

$\frac{7}{5} \times \frac{5}{9} = \frac{7}{9}$

$\frac{7}{3} \times \frac{1}{8} = \frac{7}{24}$

$\frac{4}{7} \times \frac{2}{6} = \frac{4}{21}$

$\frac{9}{10} \times \frac{7}{4} = \frac{63}{40}$

$\frac{5}{10} \times \frac{7}{7} = \frac{1}{2}$

$\frac{8}{4} \times \frac{7}{2} = \frac{7}{1}$

Multiply the fractions and reduce to simplest fraction:

$\frac{8}{3} \times \frac{6}{9} = \frac{16}{9}$ $\frac{6}{8} \times \frac{2}{5} = \frac{3}{10}$ $\frac{5}{2} \times \frac{3}{6} = \frac{5}{4}$

$\frac{6}{8} \times \frac{7}{10} = \frac{21}{40}$ $\frac{5}{6} \times \frac{2}{8} = \frac{5}{24}$ $\frac{1}{9} \times \frac{8}{4} = \frac{2}{9}$

$\frac{7}{10} \times \frac{8}{4} = \frac{7}{5}$ $\frac{8}{7} \times \frac{9}{6} = \frac{12}{7}$ $\frac{1}{4} \times \frac{3}{8} = \frac{3}{32}$

$\frac{8}{9} \times \frac{2}{4} = \frac{4}{9}$ $\frac{6}{5} \times \frac{10}{8} = \frac{3}{2}$ $\frac{10}{7} \times \frac{10}{3} = \frac{100}{21}$

Fractions, Multiplication

Unit: 31
Answers Worksheet: 3

mathbasics (1,2,3 ... to Pre-Algebra) **Target time: 3 min** **Student's time: ____ min**

Multiply the fractions and reduce to simplest fraction:

$\frac{9}{9} \times \frac{4}{10} = \frac{2}{5}$ $\frac{7}{9} \times \frac{8}{9} = \frac{56}{81}$ $\frac{5}{9} \times \frac{10}{3} = \frac{50}{27}$

$\frac{10}{9} \times \frac{1}{6} = \frac{5}{27}$ $\frac{2}{5} \times \frac{6}{3} = \frac{4}{5}$ $\frac{1}{9} \times \frac{6}{3} = \frac{2}{9}$

$\frac{1}{3} \times \frac{7}{8} = \frac{7}{24}$ $\frac{2}{6} \times \frac{7}{2} = \frac{7}{6}$ $\frac{9}{8} \times \frac{4}{7} = \frac{9}{14}$

$\frac{3}{9} \times \frac{1}{8} = \frac{1}{24}$ $\frac{10}{3} \times \frac{4}{2} = \frac{20}{3}$ $\frac{6}{8} \times \frac{7}{4} = \frac{21}{16}$

Solve the following problems and reduce to simplest fraction:

$\frac{7}{3} \div \frac{4}{10} = \frac{35}{6}$ $\frac{4}{9} \div \frac{3}{7} = \frac{28}{27}$ $\frac{2}{10} \div \frac{5}{2} = \frac{2}{25}$

$\frac{10}{8} \div \frac{10}{4} = \frac{1}{2}$ $\frac{9}{7} \div \frac{6}{9} = \frac{27}{14}$ $\frac{2}{4} \div \frac{7}{3} = \frac{3}{14}$

$\frac{7}{5} \div \frac{5}{9} = \frac{63}{25}$ $\frac{7}{5} \div \frac{8}{7} = \frac{49}{40}$ $\frac{3}{2} \div \frac{6}{2} = \frac{1}{2}$

$\frac{3}{5} \div \frac{9}{5} = \frac{1}{3}$ $\frac{7}{8} \div \frac{10}{4} = \frac{7}{20}$ $\frac{7}{4} \div \frac{4}{6} = \frac{21}{8}$

Fractions, Division

Unit: 32
Answers Worksheet: 2

mathbasics (1,2,3 ... to Pre-Algebra)   **Target time: 3 min   Student's time: ____ min**

Solve the following problems and reduce to simplest fraction:

$\frac{8}{3} \div \frac{10}{6} = \frac{8}{5}$ $\frac{10}{10} \div \frac{7}{6} = \frac{6}{7}$ $\frac{2}{6} \div \frac{10}{10} = \frac{1}{3}$

$\frac{6}{5} \div \frac{7}{9} = \frac{54}{35}$ $\frac{8}{6} \div \frac{10}{4} = \frac{8}{15}$ $\frac{2}{8} \div \frac{8}{7} = \frac{7}{32}$

$\frac{6}{10} \div \frac{6}{9} = \frac{9}{10}$ $\frac{4}{6} \div \frac{3}{10} = \frac{20}{9}$ $\frac{7}{6} \div \frac{10}{7} = \frac{49}{60}$

$\frac{4}{6} \div \frac{3}{6} = \frac{4}{3}$ $\frac{9}{2} \div \frac{6}{3} = \frac{9}{4}$ $\frac{3}{9} \div \frac{3}{8} = \frac{8}{9}$

Fractions, Division

Unit: 32
Answers Worksheet: 3

mathbasics (1,2,3 ... to Pre-Algebra) Target time: 3 min Student's time: ____ min

Solve the following problems and reduce to simplest fraction:

$\frac{9}{5} \div \frac{2}{4} = \frac{18}{5}$ $\frac{4}{8} \div \frac{3}{7} = \frac{7}{6}$ $\frac{7}{3} \div \frac{2}{10} = \frac{35}{3}$

$\frac{5}{5} \div \frac{10}{4} = \frac{2}{5}$ $\frac{2}{6} \div \frac{5}{2} = \frac{2}{15}$ $\frac{5}{4} \div \frac{9}{3} = \frac{5}{12}$

$\frac{4}{7} \div \frac{8}{4} = \frac{2}{7}$ $\frac{5}{3} \div \frac{3}{6} = \frac{10}{3}$ $\frac{10}{5} \div \frac{10}{6} = \frac{6}{5}$

$\frac{6}{2} \div \frac{8}{8} = \frac{3}{1}$ $\frac{4}{9} \div \frac{2}{10} = \frac{20}{9}$ $\frac{8}{8} \div \frac{5}{10} = \frac{2}{1}$

Simplification of Fractions

Unit: 33
Answers Worksheet: 1

mathbasics (1,2,3 ... to Pre-Algebra)   **Target time: 11 min   Student's time: ____ min**

Solve each problem and reduce to lowest simple fraction:

$\frac{12}{5} \times \frac{15}{2} \div \frac{5}{2} = \frac{36}{5}$

$\frac{6}{60} \times \frac{24}{4} \times \frac{2}{25} = \frac{6}{125}$

$\frac{9}{3} \div \frac{375}{30} \div \frac{100}{225} = \frac{27}{50}$

$\frac{3}{5} \times \frac{6}{4} \div \frac{18}{5} = \frac{1}{4}$

$\frac{3}{30} \div \frac{375}{18} \div \frac{4}{2} = \frac{3}{1250}$

$\frac{45}{40} \div \frac{5}{6} \times \frac{5}{45} = \frac{3}{20}$

$\frac{5}{15} \div \frac{45}{30} \div \frac{30}{135} = \frac{1}{1}$

$\frac{90}{9} \div \frac{60}{9} \times \frac{4}{125} = \frac{6}{125}$

$\frac{3}{3} \times \frac{12}{4} \times \frac{20}{15} = \frac{4}{1}$

$\frac{60}{9} \div \frac{30}{10} \div \frac{30}{3} = \frac{2}{9}$

mathbasics (1,2,3 ... to Pre-Algebra)   **Target time: 11 min   Student's time: ____ min**

Solve each problem and reduce to lowest simple fraction:

$\frac{4}{5} \div \frac{30}{6} \div \frac{10}{15} = \frac{6}{25}$

$\frac{90}{12} \times \frac{3}{2} \times \frac{3}{150} = \frac{9}{40}$

$\frac{100}{90} \times \frac{25}{225} \div \frac{15}{15} = \frac{10}{81}$

$\frac{30}{15} \div \frac{50}{15} \div \frac{10}{2} = \frac{3}{25}$

$\frac{5}{150} \times \frac{30}{36} \times \frac{2}{8} = \frac{1}{144}$

$\frac{135}{90} \times \frac{12}{15} \times \frac{12}{27} = \frac{8}{15}$

$\frac{3}{5} \times \frac{3}{100} \times \frac{30}{6} = \frac{9}{100}$

$\frac{75}{150} \times \frac{9}{150} \div \frac{2}{225} = \frac{27}{8}$

$\frac{4}{250} \times \frac{100}{9} \times \frac{24}{18} = \frac{32}{135}$

$\frac{15}{36} \times \frac{25}{12} \div \frac{12}{60} = \frac{625}{144}$

Simplification of Fractions

Unit: 33

Answers Worksheet: 3

(1,2,3 ... to Pre-Algebra)

Target time: 11 min **Student's time: ____ min**

Solve each problem and reduce to lowest simple fraction:

$\frac{90}{225} \div \frac{3}{90} \div \frac{20}{75} = \frac{45}{1}$

$\frac{100}{24} \times \frac{90}{2} \div \frac{3}{45} = \frac{5625}{2}$

$\frac{30}{15} \div \frac{75}{15} \div \frac{5}{6} = \frac{12}{25}$

$\frac{20}{250} \times \frac{3}{3} \div \frac{15}{6} = \frac{4}{125}$

$\frac{24}{3} \div \frac{12}{18} \times \frac{100}{6} = \frac{200}{1}$

$\frac{2}{30} \div \frac{15}{50} \times \frac{50}{3} = \frac{100}{27}$

$\frac{6}{2} \div \frac{150}{60} \times \frac{6}{30} = \frac{6}{25}$

$\frac{150}{90} \times \frac{24}{12} \times \frac{6}{250} = \frac{2}{25}$

$\frac{5}{15} \times \frac{6}{4} \times \frac{9}{24} = \frac{3}{16}$

$\frac{3}{20} \times \frac{250}{25} \times \frac{18}{27} = \frac{1}{1}$

Complex Fractions

mathbasics (1,2,3 ... to Pre-Algebra) **Target time: 11 min** **Student's time: ____ min**

Solve the following problems:

$$\frac{\frac{7}{3} \div \frac{1}{2} \times 3}{\frac{1}{5} \div \frac{1}{2} \times 4} = \frac{35}{4}$$

$$\frac{6\frac{1}{3} + 7\frac{2}{5} + 9\frac{1}{2}}{2\frac{1}{2} \div \frac{6}{3} \div 4} = \frac{5576}{75}$$

Complex Fractions

Unit: 34
Answers Worksheet: 2

mathbasics (1,2,3 ... to Pre-Algebra) **Target time: 11 min Student's time: ____ min**

Solve the following problems:

$$\frac{\frac{4}{5} \div \frac{30}{6} \div \frac{10}{15}}{\frac{90}{12} \times \frac{3}{2} \times \frac{3}{150}} = \frac{16}{15}$$

$$\frac{\frac{8}{21} + \frac{7}{42} + \frac{3}{98}}{\frac{6}{21} - \frac{1}{4} + \frac{9}{42}} = \frac{2380}{1029}$$

Solve the following problems:

$$\frac{\frac{1}{2} + \frac{2}{3} + \frac{3}{5}}{\frac{3}{5} + \frac{2}{3} - \frac{1}{2}} = \frac{53}{23}$$

$$\frac{\frac{8}{3} \div \frac{4}{6} \div \frac{9}{5}}{\frac{7}{3} + \frac{8}{2} - \frac{2}{7}} = \frac{140}{381}$$

Which law of addition is illustrated below?

47 + (14 + 7) = (47 + 14) + 7
Associative

95 + 27 = 27 + 95
Commutative

67 + 69 = 69 + 67
Commutative

45 + (44 + 26) = (45 + 44) + 26
Associative

64 + 82 = 82 + 64
Commutative

85 + (25 + 92) = (85 + 25) + 92
Associative

(71 + 83) + 58 = 71 + (83 + 58)
Associative

(92 + 14) + 6 = 92 + (14 + 6)
Associative

Use the indicated law to rewrite the expression:

Commutative: 30 + 34
30 + 34 = 34 + 30

Commutative: 47 + 76
47 + 76 = 76 + 47

Commutative: 22 + 76
22 + 76 = 76 + 22

Associative: (56 + 82) + 42
(56 + 82) + 42 = 56 + (82 + 42)

Associative: (3 + 98) + 78
(3 + 98) + 78 = 3 + (98 + 78)

Commutative: 97 + 18
97 + 18 = 18 + 97

Associative: 29 + (76 + 32)
29 + (76 + 32) = (29 + 76) + 32

Associative: (92 + 14) + 52
(92 + 14) + 52 = 92 + (14 + 52)

Laws of Addition

Unit: 35
Answers Worksheet: 2

mathbasics (1,2,3 ... to Pre-Algebra) Target time: 2 min Student's time: ____ min

Which law of addition is illustrated below?

$(57 + 56) + 59 = 57 + (56 + 59)$
Associative

$39 + (15 + 39) = (39 + 15) + 39$
Associative

$(40 + 96) + 38 = 40 + (96 + 38)$
Associative

$42 + 18 = 18 + 42$
Commutative

$38 + 85 = 85 + 38$
Commutative

$39 + 55 = 55 + 39$
Commutative

$55 + 90 = 90 + 55$
Commutative

$(11 + 7) + 82 = 11 + (7 + 82)$
Associative

Use the indicated law to rewrite the expression:

Associative: $32 + (28 + 10)$
$32 + (28 + 10) = (32 + 28) + 10$

Commutative: $15 + 61$
$15 + 61 = 61 + 15$

Commutative: $69 + 34$
$69 + 34 = 34 + 69$

Commutative: $80 + 66$
$80 + 66 = 66 + 80$

Commutative: $72 + 89$
$72 + 89 = 89 + 72$

Commutative: $63 + 28$
$63 + 28 = 28 + 63$

Commutative: $31 + 34$
$31 + 34 = 34 + 31$

Associative: $(40 + 60) + 47$
$(40 + 60) + 47 = 40 + (60 + 47)$

Laws of Addition

Unit: 35
Answers Worksheet: 3

mathbasics (1,2,3 ... to Pre-Algebra) **Target time: 2 min** **Student's time: ____ min**

Which law of addition is illustrated below?

67 + 56 = 56 + 67
Commutative

(98 + 88) + 63 = 98 + (88 + 63)
Associative

36 + (60 + 91) = (36 + 60) + 91
Associative

(45 + 56) + 52 = 45 + (56 + 52)
Associative

(46 + 40) + 78 = 46 + (40 + 78)
Associative

31 + 2 = 2 + 31
Commutative

39 + 7 = 7 + 39
Commutative

(96 + 79) + 76 = 96 + (79 + 76)
Associative

Use the indicated law to rewrite the expression:

Associative: 42 + (54 + 59)
42 + (54 + 59) = (42 + 54) + 59

Commutative: 51 + 68
51 + 68 = 68 + 51

Commutative: 92 + 69
92 + 69 = 69 + 92

Associative: (53 + 84) + 59
(53 + 84) + 59 = 53 + (84 + 59)

Associative: (88 + 1) + 89
(88 + 1) + 89 = 88 + (1 + 89)

Associative: 61 + (69 + 87)
61 + (69 + 87) = (61 + 69) + 87

Associative: 13 + (56 + 97)
13 + (56 + 97) = (13 + 56) + 97

Commutative: 64 + 60
64 + 60 = 60 + 64

Laws of Multiplication

Unit: 36
Answers Worksheet: 1

mathbasics (1,2,3 ... to Pre-Algebra) **Target time: 3 min Student's time: ____ min**

Which law of multiplication is illustrated below?

$81 \times 79 = 79 \times 81$
Commutative

$21 \times (23 \times 57) = (21 \times 23) \times 57$
Associative

$26 \times (92 + 98) = (26 \times 92) + (26 \times 98)$
Distributive

$39 \times (5 \times 74) = (39 \times 5) \times 74$
Associative

$43 \times (80 + 99) = (43 \times 80) + (43 \times 99)$
Distributive

$(82 \times 86) \times 54 = 82 \times (86 \times 54)$
Associative

$56 \times 1 = 1 \times 56$
Commutative

$53 \times (15 + 7) = (53 \times 15) + (53 \times 7)$
Distributive

Use the indicated law to rewrite the expression:

Associative: $92 \times (93 \times 90)$
$92 \times (93 \times 90) = (92 \times 93) \times 90$

Distributive: $57 \times (71 + 30)$
$57 \times (71 + 30) = (57 \times 71) + (57 \times 30)$

Commutative: 84×91
$84 \times 91 = 91 \times 84$

Distributive: $77 \times (85 + 13)$
$77 \times (85 + 13) = (77 \times 85) + (77 \times 13)$

Associative: $98 \times (74 \times 48)$
$98 \times (74 \times 48) = (98 \times 74) \times 48$

Commutative: 41×94
$41 \times 94 = 94 \times 41$

Distributive: $73 \times (23 + 73)$
$73 \times (23 + 73) = (73 \times 23) + (73 \times 73)$

Associative: $(19 \times 13) \times 57$
$(19 \times 13) \times 57 = 19 \times (13 \times 57)$

Laws of Multiplication

Unit: 36
Answers Worksheet: 2

mathbasics (1,2,3 ... to Pre-Algebra) **Target time: 3 min Student's time: ____ min**

Which law of multiplication is illustrated below?

$5 \times 25 = 25 \times 5$
Commutative

$89 \times (48 + 35) = (89 \times 48) + (89 \times 35)$
Distributive

$(30 \times 29) \times 99 = 30 \times (29 \times 99)$
Associative

$76 \times 7 = 7 \times 76$
Commutative

$(20 \times 13) \times 75 = 20 \times (13 \times 75)$
Associative

$58 \times 47 = 47 \times 58$
Commutative

$35 \times 19 = 19 \times 35$
Commutative

$31 \times 65 = 65 \times 31$
Commutative

Use the indicated law to rewrite the expression:

Distributive: $44 \times (37 + 5)$
$44 \times (37 + 5) = (44 \times 37) + (44 \times 5)$

Associative: $(79 \times 32) \times 48$
$(79 \times 32) \times 48 = 79 \times (32 \times 48)$

Commutative: 70×5
$70 \times 5 = 5 \times 70$

Associative: $(18 \times 62) \times 31$
$(18 \times 62) \times 31 = 18 \times (62 \times 31)$

Commutative: 55×77
$55 \times 77 = 77 \times 55$

Distributive: $56 \times (74 + 20)$
$56 \times (74 + 20) = (56 \times 74) + (56 \times 20)$

Commutative: 70×33
$70 \times 33 = 33 \times 70$

Distributive: $77 \times (38 + 29)$
$77 \times (38 + 29) = (77 \times 38) + (77 \times 29)$

Which law of multiplication is illustrated below?

(94 × 59) × 23 = 94 × (59 × 23)
Associative

36 × (34 + 19) = (36 × 34) + (36 × 19)
Distributive

29 × 83 = 83 × 29
Commutative

30 × (89 + 19) = (30 × 89) + (30 × 19)
Distributive

88 × 36 = 36 × 88
Commutative

43 × (82 × 9) = (43 × 82) × 9
Associative

43 × (33 + 72) = (43 × 33) + (43 × 72)
Distributive

84 × 68 = 68 × 84
Commutative

Use the indicated law to rewrite the expression:

Distributive: 62 × (83 + 94)
62 × (83 + 94) = (62 × 83) + (62 × 94)

Commutative: 42 × 44
42 × 44 = 44 × 42

Associative: 51 × (15 × 71)
51 × (15 × 71) = (51 × 15) × 71

Commutative: 14 × 29
14 × 29 = 29 × 14

Commutative: 61 × 58
61 × 58 = 58 × 61

Commutative: 40 × 97
40 × 97 = 97 × 40

Associative: 59 × (65 × 49)
59 × (65 × 49) = (59 × 65) × 49

Associative: (12 × 81) × 73
(12 × 81) × 73 = 12 × (81 × 73)

Solve the following problems:

$2 + 7 \times 3 = 23$

$(6 + 9) - 15 + 3 = 3$

$72 \div 3^2 - 2 = 6$

$27 \times 3 - 2^2 + (9 - 8) = 78$

$(2 + 5) \times 6^2 + 9 \div 3 \times 0 = 252$

$(3 \times 2) + 5 + 3^3 \times 9 = 254$

Order of Operations

Unit: 37
Answers Worksheet: 2

mathbasics (1,2,3 ... to Pre-Algebra) **Target time: 7 min** **Student's time: ____ min**

Solve the following problems:

$5 + 7 \times 3 = 26$

$9 \times 9 + 6 - 2 = 85$

$6^2 + 7 \times 2 \times 9 - 5^3 = 37$

$6 + (7 - 2) + 8^2 - 35 = 40$

$9 \times 9 - 64 + 3^2 \div 3 = 20$

$7 + 2^{12} \times 2 - 5600 = 2599$

Order of Operations

Unit: 37
Answers Worksheet: 3

mathbasics (1,2,3 ... to Pre-Algebra) **Target time: 7 min** **Student's time: ____ min**

Solve the following problems:

$630 - 7^2 \times (3 + 4) = 287$

$5 \times 10 + 7^2 \times (15 \div 5) = 197$

$3 \times 7 + 2^5 + 9 - 55 = 7$

$3^3 \times (9 + 7^2 - 2) - 20 \times 0 = 1512$

$(6 + 2) \times 4 + 10^2 - 3 = 129$

$3 \times (7 + 2^2) - 9 \times 2 = 15$

Write the following decimals in words:

0.45: forty-five hundredths

0.00002: two hundred-thousandths

0.000013: thirteen millionths

0.00018: eighteen hundred-thousandths

0.000008: eight millionths

0.09: nine hundredths

Write out each of the following as decimal numbers:

fifty-one hundred-thousandths: 0.00051

four ten-thousandths: 0.0004

forty-five ten-thousandths: 0.0045

seventy-nine ten-thousandths: 0.0079

eighteen hundred-thousandths: 0.00018

Write the following decimals in words:

0.018: eighteen thousandths

0.0005: five ten-thousandths

0.00034: thirty-four hundred-thousandths

0.002: two thousandths

0.00005: five hundred-thousandths

0.000058: fifty-eight millionths

Write out each of the following as decimal numbers:

twenty-six hundredths: 0.26

six hundred-thousandths: 0.00006

one ten-thousandths: 0.0001

ninety-seven hundred-thousandths: 0.00097

two hundred-thousandths: 0.00002

Write the following decimals in words:

0.0003: three ten-thousandths

0.00004: forty millionths

0.048: forty-eight thousandths

0.07: seven hundredths

0.0001: one ten-thousandths

0.09: nine hundredths

Write out each of the following as decimal numbers:

eight ten-thousandths: 0.0008

nine tenths: 0.9

ninety-nine hundredths: 0.99

eighty-one ten-thousandths: 0.0081

seventy-fiv millionthse: 0.000075

Addition of Decimals

Unit: 39

Answers Worksheet: 1

mathbasics (1,2,3 ... to Pre-Algebra) **Target time: 11 min Student's time: ____ min**

Solve the following problems:

89.66 + 767.9 =

857.56

2.58 + 154.076023 =

156.656023

8.9 + 2.089 =

10.989

952.6236 + 49.06 =

1001.6836

9.18 + 9.2 =

18.38

9.59 + 607.8 =

617.39

73.1684 + 4.5596 + 2.8237 =

80.5517

14.5996 + 5.05061 + 19.1 =

38.75021

65.11 + 561 + 92.406 =

718.516

8.2053 + 74.5 + 1.1034 =

83.8087

4.632 + 56.834918 + 692.083 + 48.638 = 802.187918

11.558 + 41.57 + 588.520065 + 765.54 = 1407.188065

Target time: 11 min **Student's time: ____ min**

Solve the following problems:

1.96 + 288.08 =

290.04

3.32 + 372.766365 =

376.086365

0.22459 + 584.911=

585.13559

541.206 + 64.2963 =

605.5023

78.997805 + 9.273461=

88.271266

76.738347 + 55.717261=

132.455608

501.0729 + 7.094 + 289.203601 =

797.370501

933.9871 + 47.565 + 5.87147=

987.42357

103.31319 + 44.3 + 3 =

150.61319

20.955 + 0.865 + 480.35 =

502.17

652.152 + 90.57 + 72.482097 + 471.9 = 1287.104097

457.8 + 482.443313 + 92.225 + 266.0394 = 1298.507713

Addition of Decimals

Unit: 39
Answers Worksheet: 3

mathbasics (1,2,3 ... to Pre-Algebra) **Target time: 11 min Student's time: ____ min**

Solve the following problems:

67.851948 + 632.11744 =
699.969388

15.5811 + 74.933 =
90.5141

268.342+910.1571=
1178.4991

641.732995 + 529.72 =
1171.452995

6.59 + 96.72508 =
103.31508

6.381 + 85.7 =
92.081

633.404926 + 337.710264 + 75.65109 =
1046.76628

23.362 + 4.36152+98.03517=
125.75869

0.32177 + 4.7 + 708.14 =
713.16177

3.6 + 572.02 + 8.2507 =
583.8707

58.871 + 4.582 + 290.3 + 7.36455 = 361.11755

3.7 + 62.29 + 67.53583 + 467.2216 = 600.74743

Target time: 12 min Student's time: ____ min

Solve the following problems:

7224.50175 - 111.1 = 7113.40175

1689.9049 - 842.88096 = 847.02394

4942.1 - 574 = 4368.1

4142.1592 - 443.7866 = 3698.3726

1.7522 - 1.381691 = 0.370509

7723.77 - 397.68916 = 7326.08084

149.729646 - 2.777338 = 146.952308

9881.978 - 970.94 = 8911.038

7979.64 - 788.21357 = 7191.42643

8225.69862 - 51.090485 = 8174.608135

5233.459973 - 835.04 = 4398.419973

727.1 - 7.07934 = 720.02066

1806.34133 - 74.578293 = 1731.763037

4178.756 - 64.51405 = 4114.24195

7483.797063-19.59772=7464.199343

Subtraction of Decimals

Unit: 40
Answers Worksheet: 2

mathbasics (1,2,3 ... to Pre-Algebra) Target time: 12 min Student's time: ____ min

Solve the following problems:

5344.9 - 137.9454 = 5206.9546

408.57695 - 191.5 = 217.07695

72.2 - 18.272632 = 53.927368

9.3797 - 4.551 = 4.8287

9551.0505 - 91.636764 = 9459.413736

5169.933932 - 26.8 = 5143.133932

3.740921 - 3.15274 = 0.588181

673.586 - 89.79847 = 583.78753

4084.3183 - 9.762 = 4074.5563

407.267 - 3.787881 = 403.479119

8.766453 - 6.214013 = 2.55244

616.057 - 24.038972 = 592.018028

642.289567 - 126.559 = 515.730567

623.4189 - 478.656 = 144.7629

6500.011 - 862.83 = 5637.181

Subtraction of Decimals

Unit: 40
Answers Worksheet: 3

mathbasics (1,2,3 ... to Pre-Algebra) **Target time: 12 min** **Student's time: ____ min**

Solve the following problems:

506.63 - 280.4961 = 226.1339

4787.37 - 29 = 4758.37

5589.367 - 738.1727 = 4851.1943

2495.16 - 550.4 = 1944.76

4.3403 - 2.976217 = 1.364083

2114.958 - 56.19 = 2058.768

8970.3 - 592.350273 = 8377.949727

7.324109 - 2.47 = 4.854109

62.666191 - 32.3506 = 30.315591

569.363 - 37.36 = 532.003

7294.753 - 45.6298 = 7249.1232

7516.196 - 583.1 = 6933.096

205.6469 - 194.95 = 10.6969

45.73 - 9.9437 = 35.7863

8628.8106 - 915.101 = 7713.7096

Multiplication of Decimals

Unit: 41

Answers Worksheet: 1

mathbasics (1,2,3 ... to Pre-Algebra) **Target time: 10 min** **Student's time: ____ min**

Solve the following problems:

$84.01 \times 5.72 = 480.5372$

$0.04 \times 80.4 = 3.216$

$42.4 \times 6.3 = 267.12$

$73.67 \times 27.2 = 2003.824$

$811.4 \times 52.49 = 42590.386$

$47.1 \times 6.09 = 286.839$

$17.26 \times 0.3 = 5.178$

$543.11 \times 74.23 = 40315.0553$

$8.8 \times 29.6 = 260.48$

$3.81 \times 33.99 = 129.5019$

$51.01 \times 79.67 = 4063.9667$

$71.08 \times 97.4 = 6923.192$

Multiplication of Decimals

Target time: 10 min Student's time: ____ min

Solve the following problems:

$644.59 \times 89.8 = 57884.182$

$19.9 \times 0.9 = 17.91$

$6.02 \times 61.1 = 367.822$

$659.6 \times 4.59 = 3027.564$

$328.9 \times 1 = 328.9$

$597.13 \times 3.33 = 1988.4429$

$75.3 \times 67.7 = 5097.81$

$0 \times 40.38 = 0$

$342.92 \times 97.4 = 33400.408$

$9 \times 74.4 = 669.6$

$3.6 \times 67.09 = 241.524$

$1.9 \times 64.8 = 123.12$

Multiplication of Decimals

Unit: 41
Answers Worksheet: 3

mathbasics (1,2,3 ... to Pre-Algebra) **Target time: 10 min Student's time: ____ min**

Solve the following problems:

$319.8 \times 6.7 = 2142.66$

$6.9 \times 56.16 = 387.504$

$5.4 \times 17.75 = 95.85$

$94.19 \times 24.55 = 2312.3645$

$17.64 \times 4.49 = 79.2036$

$591.9 \times 8.68 = 5137.692$

$92.2 \times 53.39 = 4922.558$

$793.68 \times 32.1 = 25477.128$

$7.6 \times 3.93 = 29.868$

$173.75 \times 32.96 = 5726.8$

$4.2 \times 25.04 = 105.168$

$6.81 \times 2.88 = 19.6128$

Division of Decimals

Unit: 42

Answers Worksheet: 1

mathbasics (1,2,3 ... to Pre-Algebra) **Target time: 13 min** **Student's time: min**

Solve the following problems:

33÷2.4 = 13.75 105÷6=17.5 2.1 ÷ 2.4 = 0.875

0.4÷0.16 = 2.5 0.5 ÷ 2 = 0.25 5.4÷ 1.35 = 4

13.5÷ 7.5 = 1.8 18÷ 48 = 0.375 1.5 ÷ 1.2 = 1.25

63÷ 36 = 1.75 64.8÷4.05 = 16 1.04÷ 0.64 = 1.625

Division of Decimals

Unit: 42
Answers Worksheet: 2

mathbasics (1,2,3 ... to Pre-Algebra) **Target time: 13 min** **Student's time:** **min**

Solve the following problems:

$0.12 \div 0.16 = 0.75$ $2.7 \div 24 = 0.1125$ $4 \div 25 = 0.16$

$31.5 \div 36 = 0.875$ $40 \div 25 = 1.6$ $0.9 \div 6 = 0.15$

$12 \div 8 = 1.5$ $0.14 \div 1.6 = 0.0875$ $10 \div 2 = 5$

$5.4 \div 9 = 0.6$ $15 \div 12 = 1.25$ $1.26 \div 0.9 = 1.4$

Division of Decimals

Unit: 42

Answers Worksheet: 3

math basics (1,2,3 ... to Pre-Algebra) **Target time: 13 min** **Student's time: min**

Solve the following problems:

$0.06 \div 0.24 = 0.25$	$3 \div 4.8 = 0.625$	$0.6 \div 3 = 0.2$
$10.5 \div 6 = 1.75$	$0.04 \div 0.16 = 0.25$	$6 \div 9.6 = 0.625$
$13 \div 80 = 0.1625$	$42 \div 30 = 1.4$	$67.5 \div 60 = 1.125$
$0.4 \div 0.08 = 5$	$2 \div 8 = 0.25$	$2.25 \div 11.25 = 0.2$

Converting Decimals to Fractions

Unit: 43
Answers Worksheet: 1

mathbasics (1,2,3 ... to Pre-Algebra) **Target time: 6 min** **Student's time: ____ min**

Convert the following decimals to fractions:

$949.6 = 949\frac{3}{5}$	$591.6 = 591\frac{3}{5}$	$551.375 = 551\frac{3}{8}$
$301.4 = 301\frac{2}{5}$	$634.25 = 634\frac{1}{4}$	$295.4 = 295\frac{2}{5}$
$778.5 = 778\frac{1}{2}$	$410.875 = 410\frac{7}{8}$	$146.6 = 146\frac{3}{5}$
$730.375 = 730\frac{3}{8}$	$123.375 = 123\frac{3}{8}$	$354.5 = 354\frac{1}{2}$
$209.25 = 209\frac{1}{4}$	$406.875 = 406\frac{7}{8}$	$160.4 = 160\frac{2}{5}$

Converting Decimals to Fractions

Unit: 43
Answers Worksheet: 2

mathbasics (1,2,3 ... to Pre-Algebra) **Target time: 6 min Student's time: ____ min**

Convert the following decimals to fractions:

$306.375 = 306\frac{3}{8}$ $375.875 = 375\frac{7}{8}$ $283.625 = 283\frac{5}{8}$

$541.625 = 541\frac{5}{8}$ $141.75 = 141\frac{3}{4}$ $522.75 = 522\frac{3}{4}$

$276.2 = 276\frac{1}{5}$ $498.75 = 498\frac{3}{4}$ $938.8 = 938\frac{4}{5}$

$358.125 = 358\frac{1}{8}$ $752.6 = 752\frac{3}{5}$ $340.8 = 340\frac{4}{5}$

$479.6 = 479\frac{3}{5}$ $919.625 = 919\frac{5}{8}$ $33.875 = 33\frac{7}{8}$

Converting Decimals to Fractions

Unit: 43
Answers Worksheet: 3

mathbasics (1,2,3 ... to Pre-Algebra) **Target time: 6 min** **Student's time: ____ min**

Convert the following decimals to fractions:

$953.5 = 953\frac{1}{2}$	$749.125 = 749\frac{1}{8}$	$69.2 = 69\frac{1}{5}$
$893.125 = 893\frac{1}{8}$	$304.8 = 304\frac{4}{5}$	$435.875 = 435\frac{7}{8}$
$999.75 = 999\frac{3}{4}$	$58.8 = 58\frac{4}{5}$	$936.5 = 936\frac{1}{2}$
$429.75 = 429\frac{3}{4}$	$730.6 = 730\frac{3}{5}$	$709.625 = 709\frac{5}{8}$
$470.375 = 470\frac{3}{8}$	$324.125 = 324\frac{1}{8}$	$609.625 = 609\frac{5}{8}$

Convert the following decimals to fractions:

$\frac{225}{125} = 1.8$ $\frac{625}{1000} = 0.625$ $\frac{36}{20} = 1.8$

$\frac{750}{1200} = 0.625$ $\frac{225}{360} = 0.625$ $\frac{125}{100} = 1.25$

$\frac{60}{48} = 1.25$ $\frac{54}{270} = 0.2$ $\frac{90}{240} = 0.375$

$\frac{36}{45} = 0.8$ $\frac{18}{144} = 0.125$ $\frac{63}{45} = 1.4$

$\frac{7.277}{1000} = 0.007277$ $\frac{69.45}{10000} = 0.006945$ $\frac{663.8}{10000} = 0.06638$

Converting Fractions to Decimals

Unit: 44
Answers Worksheet: 2

mathbasics (1,2,3 ... to Pre-Algebra) **Target time: 20 min** **Student's time: ____ min**

Convert the following decimals to fractions:

$\frac{20}{16} = 1.25$

$\frac{70}{80} = 0.875$

$\frac{450}{250} = 1.8$

$\frac{60}{96} = 0.625$

$\frac{375}{625} = 0.6$

$\frac{100}{400} = 0.25$

$\frac{60}{80} = 0.75$

$\frac{200}{125} = 1.6$

$\frac{468}{288} = 1.625$

$\frac{12}{20} = 0.6$

$\frac{132}{96} = 1.375$

$\frac{54}{135} = 0.4$

$\frac{920.4}{10} = 92.04$

$\frac{342.9}{10000} = 0.03429$

$\frac{79.67}{1000} = 0.07967$

Converting Fractions to Decimals

Unit: 44
Answers Worksheet: 3

mathbasics (1,2,3 ... to Pre-Algebra) **Target time: 20 min** **Student's time: ____ min**

Convert the following decimals to fractions:

$\frac{810}{450} = 1.8$

$\frac{420}{240} = 1.75$

$\frac{132}{96} = 1.375$

$\frac{60}{96} = 0.625$

$\frac{24}{64} = 0.375$

$\frac{100}{80} = 1.25$

$\frac{54}{36} = 1.5$

$\frac{45}{25} = 1.8$

$\frac{40}{64} = 0.625$

$\frac{45}{360} = 0.125$

$\frac{702}{432} = 1.625$

$\frac{405}{540} = 0.75$

$\frac{10.19}{100} = 0.1019$

$\frac{199.7}{1000} = 0.1997$

$\frac{8.183}{1000} = 0.008183$

Percents and Fractions

Unit: 45
Answers Worksheet: 1

mathbasics (1,2,3 ... to Pre-Algebra) **Target time: 5 min Student's time: ____ min**

Convert the following percents to a fraction (reduced to lowest terms):

160% = 8/5 162.5% = 13/8 112.5% = 9/8

112.5% = 9/8 25% = 1/4 12.5% = 1/8

140% = 7/5 137.5% = 11/8 175% = 7/4

Convet the following fractions to a percent:

$\frac{63}{36}$ = 175% $\frac{28}{32}$ = 87.5% $\frac{6}{30}$ = 20%

$\frac{24}{60}$ = 40% $\frac{264}{192}$ = 137.5% $\frac{14}{10}$ = 140%

$\frac{12}{16}$ = 75% $\frac{54}{45}$ = 120% $\frac{3}{6}$ = 50%

mathbasics (1,2,3 ... to Pre-Algebra) **Target time: 5 min** **Student's time: ____ min**

Convert the following percents to a fraction (reduced to lowest terms):

120% = 6/5 140% = 7/5 80% = 4/5

12.5% = 1/8 180% = 9/5 180% = 9/5

25% = 1/4 112.5% = 9/8 60% = 3/5

Convet the following fractions to a percent:

$\frac{6}{24}$ = 25% $\frac{132}{96}$ = 137.5% $\frac{54}{90}$ = 60%

$\frac{9}{12}$ = 75% $\frac{24}{60}$ = 40% $\frac{72}{60}$ = 120%

$\frac{30}{48}$ = 62.5% $\frac{42}{24}$ = 175% $\frac{9}{36}$ = 25%

Convert the following percents to a fraction (reduced to lowest terms):

137.5% = 11/8 87.5% = 7/8 150% = 3/2

40% = 2/5 20% = 1/5 12.5% = 1/8

12.5% = 1/8 87.5% = 7/8 62.5% = 5/8

Convet the following fractions to a percent:

$\frac{81}{108}$ = 75% $\frac{324}{180}$ = 180% $\frac{162}{144}$ = 112.5%

$\frac{54}{48}$ = 112.5% $\frac{54}{36}$ = 150% $\frac{14}{10}$ = 140%

$\frac{21}{12}$ = 175% $\frac{6}{30}$ = 20% $\frac{6}{24}$ = 25%

Percents and Decimals

Unit: 46
Answers Worksheet: 1

mathbasics (1,2,3 ... to Pre-Algebra) **Target time: 5 min Student's time: ____ min**

Convert the percents to decimal:

848.2904% =	737.9254% =	4.23498% =	319.149% =
8.482904	7.379254	0.0423498	3.19149
47.8% =	456.1326% =	573.36% =	155.239408% =
0.478	4.561326	5.7336	1.55239408
968.807335% =	9.9% =	824.958939% =	73.57% =
9.68807335	0.099	8.24958939	0.7357
653.4% =	257.81% =	328.170513% =	202.52644% =
6.534	2.5781	3.28170513	2.0252644

Convert the decimals to percent:

22.0 =	5.9 =	502.6 =	964.8440 =
2200%	590%	50260%	96484.4%
0.4 =	4.6 =	9.45438 =	856.1 =
40%	460%	945.438%	85610%
651.22 =	5.392653 =	3.1931 =	776.36 =
65122%	539.2653%	319.31%	77636%
9.9 =	561.78441 =	1.511704 =	1.5089 =
990%	56178.441%	151.1704%	150.89%

Percents and Decimals

Unit: 46
Answers Worksheet: 2

mathbasics (1,2,3 ... to Pre-Algebra) Target time: 5 min Student's time: ____ min

Convert the percents to decimal:

2.7629% =	62.8013% =	98.828% =	2.22% =
0.027629	0.628013	0.98828	0.0222
566.9435% =	0.3% =	686.5697% =	4.4% =
5.669435	0.003	6.865697	0.044
866.245% =	247.1% =	658.26176% =	551.7% =
8.66245	2.471	6.5826176	5.517
3.5727% =	37.98260% =	1.874274% =	4.11874% =
0.035727	0.379826	0.01874274	0.0411874

Convert the decimals to percent:

68.951 =	55.7 =	20.229776 =	66.0 =
6895.1%	5570%	2022.9776%	6600%
55.6926 =	958.63261 =	498.740 =	89.16960 =
5569.26%	95863.261%	49874%	8916.96%
978.5167 =	786.027661 =	24.30802 =	32.677 =
97851.67%	78602.7661%	2430.802%	3267.7%
884.7759 =	9.93 =	5.829 =	874.29442 =
88477.59%	993%	582.9%	87429.442%

Percents and Decimals

Unit: 46
Answers Worksheet: 3

mathbasics (1,2,3 ... to Pre-Algebra)   Target time: 5 min   Student's time: ____ min

Convert the percents to decimal:

242.5856% =	23.3% =	19.6% =	283.56564% =
2.425856	0.233	0.196	2.8356564
42.4% =	9.95039% =	188.130% =	184.30879% =
0.424	0.0995039	1.8813	1.8430879
166.2339% =	5.1114% =	195.36293% =	5.58528% =
1.662339	0.051114	1.9536293	0.0558528
722.46% =	1.00% =	7.28% =	27.92% =
7.2246	0.01	0.0728	0.2792

Convert the decimals to percent:

264.4 =	2.3078 =	69.11663 =	3.464216 =
26440%	230.78%	6911.663%	346.4216%
11.458372 =	30.47 =	4.4095 =	1.2 =
1145.8372%	3047%	440.95%	120%
5.793546 =	6.293743 =	145.5523 =	605.8044 =
579.3546%	629.3743%	14555.23%	60580.44%
963.3 =	71.36 =	363.01328 =	641.91818 =
96330%	7136%	36301.328%	64191.818%

Percent Problems - Type A

Unit: 47
Answers Worksheet: 1

mathbasics (1,2,3 ... to Pre-Algebra) **Target time: 9 min Student's time: ____ min**

Solve the following problems:

77% of 58 = 44.66	80% of 63 = 50.4	41% of 20 = 8.2
60% of 50 = 30	20% of 79 = 15.8	17% of 2 = 0.34
15% of 699 = 104.85	60% of 792 = 475.2	5% of 779 = 38.95
5.16% of 9.2 = 0.47472	83.4% of 7.07 = 5.89638	7.64% of 6.2 = 0.47368
17.3% of 2.04 = 0.35292	2.8% of 1.9 = 0.0532	34.1% of 74.17 = 25.29197

Percent Problems - Type A

Unit: 47
Answers Worksheet: 2

mathbasics (1,2,3 ... to Pre-Algebra) **Target time: 9 min Student's time: ____ min**

Solve the following problems:

18% of 63 = 11.34	8% of 56 = 4.48	8% of 25 = 2
50% of 68 = 34	84% of 95 = 79.8	76% of 7 = 5.32
36% of 588 = 211.68	71% of 346 = 245.66	83% of 166 = 137.78
78.79% of 9.5 = 7.48505	71.58% of 17.6 = 12.59808	97.5% of 41.3 = 40.2675
4% of 27.16 = 1.0864	62.9% of 8.6 = 5.4094	33.8% of 8.43 = 2.84934

Percent Problems - Type A

mathbasics (1,2,3 ... to Pre-Algebra) Target time: 9 min Student's time: ____ min

Solve the following problems:

33% of 76 = 25.08	88% of 93 = 81.84	53% of 64 = 33.92
100% of 54 = 54	1% of 58 = 0.58	36% of 23 = 8.28
91% of 250 = 227.5	97% of 534 = 517.98	14% of 768 = 107.52
43.15% of 6.9 = 2.97735	2.92% of 61 = 1.7812	7.86% of 3.14 = 0.246804
47.77% of 14.5 = 6.92665	88.6% of 3.4 = 3.0124	45.1% of 3.7 = 1.6687

Percent Problems - Type B

Unit: 48
Answers Worksheet: 1

math basics (1,2,3 ... to Pre-Algebra) **Target time: 15 min** **Student's time: ____ min**

Solve the following problems:

12 is what percent of 96? 12.5%

42 is what percent of 30? 140%

180 is what percent of 100? 180%

4125 is what percent of 3000? 137.5%

2.52 is what percent of 1.44? 175%

5.4 is what percent of 6.75? 80%

What percent of 72 is 99? 137.5%

What percent of 72 is 90? 125%

What percent of 600 is 825? 137.5%

What percent of 180 is 252? 140%

What percent of 9 is 2.25? 25%

What percent of 0.6 is 1.08? 180%

Percent Problems - Type B

Unit: 48
Answers Worksheet: 2

mathbasics (1,2,3 ... to Pre-Algebra) **Target time: 15 min** **Student's time: ____ min**

Solve the following problems:

990 is what percent of 720? 137.5%

140 is what percent of 100? 140%

540 is what percent of 450? 120%

25 is what percent of 20? 125%

6.75 is what percent of 11.25? 60%

0.04 is what percent of 0.2? 20%

What percent of 20 is 15? 75%

What percent of 750 is 600? 80%

What percent of 1250 is 1750? 140%

What percent of 1875 is 1125? 60%

What percent of 4 is 0.5? 12.5%

What percent of 1.5 is 2.25? 150%

Percent Problems - Type B

Unit: 48
Answers Worksheet: 3

mathbasics (1,2,3 ... to Pre-Algebra) **Target time: 15 min** **Student's time: ____ min**

Solve the following problems:

525 is what percent of 375? 140%

2475 is what percent of 1800? 137.5%

6 is what percent of 4? 150%

240 is what percent of 150? 160%

6.6 is what percent of 4.8? 137.5%

5.4 is what percent of 4.8? 112.5%

What percent of 400 is 250? 62.5%

What percent of 200 is 325? 162.5%

What percent of 3000 is 4875? 162.5%

What percent of 144 is 162? 112.5%

What percent of 0.08 is 0.14? 175%

What percent of 7.2 is 0.9? 12.5%

Percent Problems - Type C

Unit: 49
Answers Worksheet: 1

mathbasics (1,2,3 ... to Pre-Algebra) **Target time: 24 min** **Student's time: ____ min**

Solve the following problems:

375 is 25% of what number? 1500

252 is 87.5% of what number? 288

60 is 75% of what number? 80

36 is 20% of what number? 180

1.8 is 62.5% of what number? 2.88

0.21 is 87.5% of what number? 0.24

12 is 12.5% of what number? 96

70 is 175% of what number? 40

75 is 150% of what number? 50

12 is 60% of what number? 20

2.4 is 80% of what number? 3

6.75 is 180% of what number? 3.75

Percent Problems - Type C

Unit: 49
Answers Worksheet: 2

mathbasics (1,2,3 ... to Pre-Algebra) **Target time: 24 min** **Student's time: ____ min**

Solve the following problems:

15 is 150% of what number? 10

20 is 40% of what number? 50

270 is 37.5% of what number? 720

120 is 75% of what number? 160

1.35 is 150% of what number? 0.9

2.4 is 160% of what number? 1.5

270 is 37.5% of what number? 720

216 is 180% of what number? 120

1080 is 160% of what number? 675

270 is 112.5% of what number? 240

0.04 is 50% of what number? 0.08

0.3 is 20% of what number? 1.5

Percent Problems - Type C

Unit: 49
Answers Worksheet: 3

mathbasics (1,2,3 ... to Pre-Algebra) **Target time: 24 min** **Student's time: ____ min**

Solve the following problems:

120 is 125% of what number? 96

126 is 175% of what number? 72

140 is 175% of what number? 80

18 is 12.5% of what number? 144

8.1 is 120% of what number? 6.75

0.54 is 120% of what number? 0.45

90 is 50% of what number? 180

90 is 20% of what number? 450

50 is 20% of what number? 250

130 is 162.5% of what number? 80

0.1 is 125% of what number? 0.08

12.15 is 180% of what number? 6.75

Solve the following problems:

1. Chris earned $45 mowing lawns. Sarah earned $50 mowing lawns. What is the ratio of Chris's money to Sarah's? Answer: 9:10

2. Jenny had 50 pencils at the beginning of the year and had 25 pencils at the end of the year. What is the ratio of pencils from the beginning of the year to the end? Answer: 2:1

3. Jimmy has 21 baseball cards and Joey has 14 baseball cards. What is the ratio of Joey's cards to Jimmy's? Answer: 2:3

4. A farm sells one ear of corn for every 3 tomatoes they sell. What is the ratio of tomatoes to corn? Answer: 3:1

5. At a restaurant, there are 4 chairs for every table. If there are 60 chairs, how many tables are there? Answer: 15

6. There are 20 students in Mr. Marshal's class. If the ratio of girls to boys is 3:1, how many girls and boys are in his class? Answer: 15 girls, 5 boys

7. The ratio for the number of chocolate bars to lollipops is 4:2. If 24 lollipops were sold, how many chocolate bars were sold? Answer: 8 lollipops, 16 chocolate bars

8. There are 30 flowers in Meg and Janet's garden. If the ratio of marigolds to roses is 2:1, how many of each flower do they have? Answer: 20 marigolds, 10 roses

9. Ms. Jone's class surveyed 90 people and found that 3:6 people preferred water to soda. How many people preferred water and how many people preferred soda? Answer: 30 water, 60 soda

10. For every 5 computers sold, 3 people buy speakers. If 27 speakers were sold, how many computers were sold? Answer: 45 computers

mathbasics (1,2,3 ... to Pre-Algebra) Target time: 5 min Student's time: ____ min

Solve the following problems:

1. Sally has 50 grew roses. Stacy grew 35 roses. What is the ratio of Stacy's roses to Sally's? Answer: 7:10

2. Kendall had $66 at the beginning of the month and had $121 at the end of the month. What is the ratio of his money from the beginning of the month to the end? Answer: 6:11

3. Carol has 56 pairs of shoes and Jill has 64 pairs of shoes. What is the ratio of Jill's shoes to Carol's? Answer: 8:7

4. An aquarium has two starfish for every 10 goldfish. What is the ratio of starfish to goldfish? Answer: 1:5

5. In an apartment complex, there are 30 apartments to ever building. If there are 150 apartments, how many buildings are there? Answer: 5 buildings

6. There were 32 pizzas sold at The Pizza Place. If the ratio of pepperoni pizza to cheese is 5:3, how many pepperoni and cheese pizzas were ordered? Answer: 20 pepperoni, 12 cheese

7. The ratio for the number of chickens to cows on a farm is 3:4. If there are 42 animals on the farm, how many of each animal are there? Answer: 18 chickens, 24 cows

8. There are 55 pairs of socks in Maggie's sock drawer. If the ratio of ankle socks to normal socks is 3:8, how many of each kind of sock is there? Answer: 15 ankle socks, 40 normal socks

9. Mrs. Jackson's class has 45 kids. If the ratio of boys to girls is 5:4, how many girls and boys are in the class? Answer: 20 girls, 25 boys

10. For every 7 customers at a store, one person gets a free gift. If 54 gifts were given out, how many customers came to the store? Answer: 378 customers

Target time: 5 min Student's time: ____ min

Solve the following problems:

1. Fran has collected 20 key chains. Felicia has collected 35 key chains. What is the ratio of Fran's roses to Felicia's? Answer: 4:7

2. Carrie had a $50 gift certificate and spent $40 dollars of it. What is the ratio of the amount of money she has left to her original amount? Answer: 1:5

3. Sam grilled 150 hotdogs at the grilling tournament and Joe grilled 75. What is the ratio of Joe's hotdogs to Sam's? Answer: 1:2

4. A dentist office gives two packets of floss for every two toothbrushes they give out. What is the ratio of floss to toothbrushes? Answer: 1:1

5. A credit card company gives one reward point for every $10 spent on their card. If a person has 450 reward points, how much money have they spent? Answer: $4500

6. There were 56 raffle tickets in a bucket. If the ratio of winners to non-winners was 2:5, how many winners and non-winners were there? Answer: 16 winners, 40 non-winners

7. A store had a HUGE closeout sale and the ratio of new customers to regulars was 5:2. If there were 567 customers, how many of each customer was there? Answer: 405 new customers, 162 regulars

8. There are 36 workers in a factory. If the ratio of the packers to machine operators in 2:4, how many of each kind of worker is there? Answer: 12 packers, 24 machine operators

9. Mr. Wheeler's class has 27 kids. If the ratio of people wearing shorts to pants is 1:2, how many people are wearing shorts and how many are wearing pants? Answer: 9 shorts, 18 pants

10. Bobby gave out 2 apples for every 9 people that went to the health fair. If 108 people came to the health fair, how many apples were given out? Answer: 24 apples

Solve the following problems:

3:9::5:?

15

2:4::7:?

14

5:?::6:42

35

9:?::3:12

36

If 3 birds lay 15 eggs, how many birds would it take to lay 40 eggs?

8 birds

If 10 people can eat 8 cakes, how many people would it take to eat 20 cakes?

25 people

If 5 boys can lift 100 pounds, how many boys would it take to lift 40 pounds?

2 boys

If 4 bees pollinate 10 flowers, how many flowers can 8 bees pollinate?

20 flowers

If 8 girl scouts sell 32 cookies, how many cookies can 15 girl scouts sell?

60 cookies

If 3 people make 12 jackets, how many jackets can 20 people make?

80 jackets

Solve the following problems:

5:40::9:?

72

2:24::3:?

36

7:?::2:36

126

12:?::16:64

48

If 2 machines make 40 toys, how many machines would it take to make 100 toys?

5 machines

If 3 people can make 18 candy bars, how many people would it take to make 60 candy bars?

10 people

If 4 horses can pull 3 carriages, how many horses would it take to pull 27 carriages?

36 horses

If 7 people washed 21 towels, how many towels can 4 people wash?

12 towels

If 24 coins are worth $25, how much is 48 coins worth?

$50

If 12 kids made 4 pies, how many pies can 90 kids make?

30 pies

Proportions

Target time: 3 min **Student's time: ____ min**

Solve the following problems:

3:9::5:?

15

15:3::?:8

40

4:?::6:12

8

9:?::3:12

36

If 3 birds lay 15 eggs, how many birds would it take to lay 40 eggs?

8 birds

If 3 people can paint 2 houses in a day, how many people would it take to paint 10 houses?

15 people

If 6 plants produce 10 vegetables, how many plants would it take to grow 25 vegetables?

15 plants

If 4 girls build 10 birdhouses, how many birdhouses can 14 girls make?

35 birdhouses

If 6 fish eat 12 worms, how many worms can 17 fish eat?

34 worms

If 2 jars of jelly fill 14 packets, how many packets can 7 jars fill?

49 packets

Signed Numbers, Addition

Unit: 52
Answers Worksheet: 1

mathbasics (1,2,3 ... to Pre-Algebra) **Target time: 3 min Student's time: ____ min**

Solve the following problems:

25 + 42 = 67	-52 + 48 = -4	28 + 40 = 68
-14 + 91 = 77	86 - 39 = 47	-10 + 29 = 19
73 + 46 = 119	50 - 45 = 5	23 + 52 = 75
-94 + 12 = -82	25 - 19 = 6	-35 - 57 = -92
-44 + 59 = 15	-36 + 100 = 64	80 - 39 = 41
87 + 90 = 177	82 + 94 = 176	-68 + 84 = 16

Signed Numbers, Addition

Unit: 52
Answers Worksheet: 2

mathbasics (1,2,3 ... to Pre-Algebra) **Target time: 3 min** **Student's time: ____ min**

Solve the following problems:

-39 - 84 = -123	-76 - 32 = -108	-26 - 71 = -97
-17 - 53 = -70	20 + 94 = 114	38 - 4 = 34
86 - 90 = - 4	-8 - 36 = - 44	-66 - 47 = -113
-58 - 72 = -130	48 + 43 = 91	39 - 2 = 37
14 + 74 = 88	6 - 37 = -31	-30 + 99 = 69
-96 + 4 = -92	94 + 49 = 143	-52 + 38 = -14

Signed Numbers, Addition

Unit: 52
Answers Worksheet: 3

mathbasics (1,2,3 ... to Pre-Algebra) **Target time: 3 min** **Student's time: ____ min**

Solve the following problems:

-61 - 84 = -145	-27 - 74 = -101	36 - 15 = 21
-55 - 71 = -126	-80 - 99 = -179	-67 + 3 = -64
34 + 8 = 42	-79 - 42 = -121	58 - 79 = -21
77 + 60 = 137	-93 - 42 = -135	60 + 83 = 143
-28 + 14 = -14	- 4 + 13 = 9	5 + 91 = 96
91 + 3 = 94	64 + 78 = 142	91 - 28 = 63

Solve the following problems:

44 - 41 = 3	-7 - (-8) = 1	-38 - (-60) = 22
-27 - 56 = -83	5 - (-45) = 50	-54 - (-62) = 8
9 - 38 = -29	63 - (-56) = 119	-28 - 59 = -87
-74 - 28 = -102	-58 - (-70) = 12	7 - (-54) = 61
-37 - (-7) = -30	-98 - 63 = -161	57 - 34 = 23
61 - (-3) = 64	74 - (-30) = 104	-59 - (-66) = 7

Signed Numbers, Subtraction

Unit: 53
Answers Worksheet: 2

mathbasics (1,2,3 ... to Pre-Algebra) **Target time: 3 min** **Student's time: ____ min**

Solve the following problems:

-25 - (-23) = -2	33 - (-28) = 61	-21 - 66 = -87
-84 - (-74) = -10	-51 - (-92) = 41	-16 - (-15) = -1
33 - (-80) = 113	3 - 12 = -9	- 43 - (-12) = -31
-71 - 95 = -166	-34 - 43 = -77	-21 - (- 45) = 24
9 - 94 = -85	-63 - (- 4) = -59	-14 - (-96) = 82
-73 - 11 = -84	37 - (-82) = 119	28 - (-72) = 100

Signed Numbers, Subtraction

Unit: 53
Answers Worksheet: 3

mathbasics (1,2,3 ... to Pre-Algebra) **Target time: 3 min** **Student's time: ____ min**

Solve the following problems:

82 - 44 = 38	-67 - 25 = -92	8 - 72 = -64
75 - (-31) = 106	-83 - (-35) = - 48	-17 - (-68) = 51
12 - (-23) = 35	-96 - (-64) = -32	-64 - 38 = -102
69 - 61 = 8	-31 - (-25) = -6	-1 - 92 = -93
62 - (-2) = 64	85 - 27 = 58	-71 - 21 = -92
-10 - 33 = - 43	-8 - 46 = -54	-73 - 40 = -113

Signed Numbers, Multiplication

Unit: 54
Answers Worksheet: 1

mathbasics (1,2,3 ... to Pre-Algebra) **Target time: 8 min Student's time: ____ min**

Solve the following problems:

$25 \times 73 = 1825$	$-28 \times 29 = -812$	$-75 \times 26 = -1950$
$-5 \times -33 = 165$	$-55 \times -89 = 4895$	$-48 \times 94 = -4512$
$19 \times -16 = -304$	$99 \times 11 = 1089$	$46 \times 14 = 644$
$-65 \times -24 = 1560$	$-27 \times 14 = -378$	$32 \times 44 = 1408$
$-56 \times 23 = -1288$	$89 \times 79 = 7031$	$84 \times -72 = -6048$
$51 \times 1 = 51$	$-61 \times -51 = 3111$	$80 \times 94 = 7520$

Target time: 8 min   Student's time: ____ min

Solve the following problems:

$-27 \times -41 = 1107$	$-37 \times -45 = 1665$	$-97 \times 69 = -6693$
$-18 \times -97 = 1746$	$94 \times 52 = 4888$	$29 \times 25 = 725$
$-9 \times 23 = -207$	$-29 \times 82 = -2378$	$23 \times -32 = -736$
$65 \times 65 = 4225$	$-68 \times -81 = 5508$	$25 \times -7 = -175$
$-31 \times -15 = 465$	$-63 \times -29 = 1827$	$-65 \times 30 = -1950$
$26 \times 92 = 2392$	$70 \times -3 = -210$	$68 \times -16 = -1088$

Solve the following problems:

-10 x 94 = -940	59 x 23 = 1357	86 x 12 = 1032
-71 x -20 = 1420	- 41 x 14 = -574	-90 x -76 = 6840
-89 x 55= - 4895	-51 x -22 = 1122	88 x -39 = -3432
-26 x -80 = 2080	65 x 64 = 4160	9 x 38 = 342
96 x 71 = 6816	-59 x 30 = -1770	-5 x - 4 = 20
-61 x - 40 = 2440	85 x -58= - 4930	-88 x 60 = -5280

Signed Numbers, Division

mathbasics (1,2,3 ... to Pre-Algebra)

Target time: 6 min Student's time: ____ min

Solve the following problems:

144 ÷ 48 = 3

-3 ÷ -1 = 3

-750 ÷ 75 = -10

-322 ÷ - 46 = 7

-910 ÷ 91 = -10

0 ÷ 78 = 0

0 ÷ 57 = 0

-282 ÷ 47 = -6

950 ÷ 95 = 10

-90 ÷ -9 = 10

160 ÷ -20 = -8

-198 ÷ -99 = 2

474 ÷ 79 = 6

864 ÷ -96 = -9

-540 ÷ -60 = 9

-56 ÷ -28 = 2

124 ÷ -62 = -2

- 445 ÷ 89 = -5

Solve the following problems:

312 ÷ -39 = -8	201 ÷ 67 = 3	200 ÷ -50= - 4
126 ÷ 14 = 9	-188 ÷ 94 = -2	-240 ÷ 48 = -5
198 ÷ 66 = 3	-580 ÷ 58 = -10	-166 ÷ -83 = 2
- 432 ÷ -72 = 6	160 ÷ 16 = 10	10 ÷ -2 = -5
- 400 ÷ 40 = -10	-66 ÷ -22 = 3	44 ÷ 11 = 4
840 ÷ -84 = -10	20 ÷ - 4 = -5	-128 ÷ 16 = -8

Signed Numbers, Division

Unit: 55
Answers Worksheet: 3

mathbasics (1,2,3 ... to Pre-Algebra) **Target time: 6 min** **Student's time: ____ min**

Solve the following problems:

-74 ÷ 37 = -2	-98 ÷ 98 = -1	250 ÷ -50 = -5
0 ÷ -84 = 0	-200 ÷ - 40 = 5	355 ÷ -71 = -5
42 ÷ -6 = -7	-837 ÷ -93 = 9	-23 ÷ 23 = -1
-36 ÷ 4 = -9	240 ÷ 48 = 5	276 ÷ 69 = 4
265 ÷ -53 = -5	140 ÷ 14 = 10	-160 ÷ - 40= 4
231 ÷ 33 = 7	-231 ÷ 33 = -7	-590 ÷ 59 = -10

Absolute Value

Unit: 56
Answers Worksheet: 1

math basics (1,2,3 ... to Pre-Algebra) **Target time: 4 min** **Student's time: ____ min**

Solve the following problems:

$\|-83\| = 83$	$\|-78\| = 78$	$\|38\| = 38$	$\|-13\| = 13$
$\|-7\| = 7$	$\|1\| = 1$	$\|-67\| = 67$	$\|-46\| = 46$
$\|-25\| = 25$	$\|48\| = 48$	$\|60\| = 60$	$\|26\| = 26$

$\|-213 \div -71\| = 3$	$\|17 + 27\| = 44$	$\|-2 + 48\| = 46$
$\|36 - 36\| = 0$	$\|3 - 64\| = 61$	$\|-59 \times 33\| = 1947$
$\|-57 + 2\| = 55$	$\|328 \div 82\| = 4$	$\|-324 \div 54\| = 6$
$\|-83 \times 55\| = 4565$	$\|89 + (-23)\| = 66$	$\|-588 \div -84\| = 7$

Absolute Value

Unit: 56
Answers Worksheet: 2

mathbasics (1,2,3 ... to Pre-Algebra) **Target time: 4 min** **Student's time: ____ min**

Solve the following problems:

$\|-42\| = 42$	$\|49\| = 49$	$\|-91\| = 91$	$\|20\| = 20$
$\|-51\| = 51$	$\|-34\| = 34$	$\|-64\| = 64$	$\|53\| = 53$
$\|-44\| = 44$	$\|75\| = 75$	$\|74\| = 74$	$\|99\| = 99$

$\|-288 \div 96\| = 3$	$\|20 - (-95)\| = 115$	$\|82 \times 41\| = 3362$
$\|236 \div -59\| = 4$	$\|-20 \times -87\| = 1740$	$\|60 + (-30)\| = 30$
$\|6 \times -28\| = 168$	$\|61 + 28\| = 89$	$\|-54 + (-62)\| = 116$
$\|-31 \times -58\| = 1798$	$\|55 + (-99)\| = 44$	$\|73 \times -50\| = 3650$

Solve the following problems:

$\lvert 51 \rvert = 51$	$\lvert -66 \rvert = 66$	$\lvert -34 \rvert = 34$	$\lvert -90 \rvert = 90$
$\lvert 94 \rvert = 94$	$\lvert 80 \rvert = 80$	$\lvert -44 \rvert = 44$	$\lvert -87 \rvert = 87$
$\lvert 15 \rvert = 15$	$\lvert -21 \rvert = 21$	$\lvert -85 \rvert = 85$	$\lvert -98 \rvert = 98$

$\lvert -124 \div 62 \rvert = 2$	$\lvert -60 \times -56 \rvert = 3360$	$\lvert 0 \div 30 \rvert = 0$
$\lvert 74 - 26 \rvert = 48$	$\lvert -38 \div -38 \rvert = 1$	$\lvert -480 \div -48 \rvert = 10$
$\lvert -22 \times -24 \rvert = 528$	$\lvert 70 + (-64) \rvert = 6$	$\lvert -36 \times 20 \rvert = 720$
$\lvert -68 - 9 \rvert = 77$	$\lvert 396 \div 66 \rvert = 6$	$\lvert 14 \times 89 \rvert = 1246$

www.ingramcontent.com/pod-product-compliance
Ingram Content Group UK Ltd.
Pitfield, Milton Keynes, MK11 3LW, UK
UKHW051510290926
15471UKWH00008B/1540

9 781613 800614